T0401571

CELL BIOLOGY RESEARCH PROGRESS

A CLOSER LOOK AT METALLOPROTEINASES

Cell Biology Research Progress

Additional books and e-books in this series can be found
on Nova's website under the Series tab.

CELL BIOLOGY RESEARCH PROGRESS

A CLOSER LOOK AT METALLOPROTEINASES

LENA GOODWIN
EDITOR

NOTICE TO THE READER

Additional color graphics may be available in the e-book version of this book.

Library of Congress Cataloging-in-Publication Data

Names: Goodwin, Lena, editor.
Title: A closer look at metalloproteinases / Lena Goodwin, editor.
Description: New York : Nova Science Publishers, [2019] | Series: Cell biology research progress | Includes bibliographical references and index. |
Identifiers: LCCN 2019044612 (print) | LCCN 2019044613 (ebook) | ISBN 9781536165173 (hardcover) | ISBN 9781536165272 (adobe pdf)
Subjects: LCSH: Metalloproteinases.
Classification: LCC QP601.7 .C56 2019 (print) | LCC QP601.7 (ebook) | DDC 572/.76--dc23
LC record available at https://lccn.loc.gov/2019044612
LC ebook record available at https://lccn.loc.gov/2019044613

Published by Nova Science Publishers, Inc. † New York

CONTENTS

PREFACE

A Closer Look at Metalloproteinases first presents an analysis of the matrix metalloproteinases system in two common neurodegenerative disorders, namely age-related macular degeneration and Alzheimer's disease.

Next, this collection aims to evaluate the expression of matrix metalloproteinase-9 in nasopharyngeal carcinoma patients. The cross-sectional analytic study enrolled 106 patients with nasopharyngeal carcinoma based on radiologic and histopathological examination. The patient never underwent radiotherapy, chemotherapy and/or a combination. Matrix metalloproteinase-9 was assessed using immunohistochemical staining of matrix metalloproteinase-9 in the anatomic pathology department.

The authors go on to discuss the matrix metalloproteinases family, their structural organization, functions and role in cardiovascular disease.

Data about the role of matrix metalloproteinases in the pathogenesis of psoriasis is summarized. In psoriasis, matrix metalloproteinases facilitate structural remodeling of the epidermis that, in turn, results in the development of psoriatic plaques.

Additionally, pregnancy is a very peculiar event in which the uterus undergoes intense morphophysiological modifications, especially related to vascular remodeling and angiogenesis, as well as trophoblast invasion, which occurs in early pregnancy period until reaching maternal spiral arteries and replacing the endothelium forming the endovascular-trophoblast. Such vascular modifications ensure the high-flow and low-pressure blood supply to the developing fetus, which in turn requires more and more space, causing the uterus to become enlarged and distended. In this context, an orchestrated regulation of matrix metalloproteinases and their endogenous inhibitors is essential for proper gestational development.

In the penultimate chapter, tissue components, matrix metallo-proteinases properties and functions, bone modelling, remodelling and resorption, repair and regeneration, and pathological bone resorption are discussed.

The concluding chapter explores the way in which protein tyrosine phosphatases are important targets that are known to play a key role in the development of chronic degenerative diseases such as obesity, diabetes, and some neurological diseases.

Chapter 1 - Extracellular matrices (ECMs) play a pivotal role in regulating the delivery of nutrients and removal of waste products between cells and their blood supply. The structural architecture and composition of ECMs is maintained by tightly coupled processes of continuous synthesis and degradation. Degradation is mediated by a family of Zn^{2+}-containing, Ca^{2+}-dependent proteolytic enzymes called the matrix metalloproteinases (MMPs). These enzymes are released as latent proteins (pro-MMPs) and on activation (following the catalytic removal of a small peptide) are capable of degrading most components of an ECM. The catalytic activity of MMPs is modulated by the presence of tissue inhibitors of MMPs (TIMPs) and current opinion holds that the ratio of TIMPs/MMPs determines the relative rate of degradation. Thus, elevated ratios are thought to compromise degradation leading to the accumulation of abnormal ECM material (as in Bruch's membrane of the eye), whilst diminished ratios are thought to lead to excessive ECM degradation allowing the spread of cancer cells, and progression of angiogenesis. However, this simplistic control system for MMP activity now appears to be far more complex and has implications for our understanding of disease mechanisms involving the MMP system. The MMP species tend to undergo covalent modification leading to homo- and hetero-dimerization and non-covalent aggregation resulting in the formation of very large macromolecular weight MMP complexes (LMMC). In addition, the various MMP species also show a free-bound compartmentalisation. The net result of these changes is to reduce the availability of the latent forms of MMPs for the activation process. An assessment of the degradation potential of the MMP system in any tissue must therefore take into account the degree of sequestration of the latent MMP species, a protocol that has not previously been addressed. Based on the complexities already described, the authors will present an analysis of the MMP system in two common neurodegenerative disorders, namely age-related macular degeneration (AMD) and Alzheimer's disease (AD).

Chapter 2 - Nasopharyngeal carcinoma (NPC) is one of the most predominant cancers in Southern China and Southeast Asia. Matrix metalloproteinase (MMP)-9 has played several crucial roles in carcinogeneses, such as tumor invasion, metastases, and tumor vascularization. Extracellular matrix degradation has been linked to aggressive tumor invasion into the surrounding tissue as well as vascular or lymphatic vessels. This study aims to evaluate the expression of MMP-9 in nasopharyngeal carcinoma patients. The cross-sectional analytic study enrolled 106 patients with nasopharyngeal carcinoma based on radiologic and histopathological examination. The patient never underwent radiotherapy, chemotherapy and/or a combination. MMP-9 was assessed using immunohistochemical staining of MMP-9 in the Anatomic Pathology Department. In this study, most NPC patients were found in the age group of 41-60 as much as 72 patients (56.6%). Most NPC patients were males as much as 90 patients (69.8%). The histopathology showed the most overexpression was found in non-keratinizing squamous cell carcinoma (80.5%), Primary Tumor T3 (35.4%), Nodes N3(56.1%), and stage IV (70.7%). Through an analysis study using Fisher's exact test, it was found p value < 0.001, which showed a significant correlation between MMP-9 overexpression to the clinical stage of NPC. It suggests the evidence that MMP-9 must be unveiled its potency as a biomarker of nasopharyngeal caricnoma. The high expression of MMP-9 in nasopharyngeal carcinoma patients supported the evidence of its role in the disease and so the potentiation of the inhibitor of MMP-9 as targeted therapy of nasopharyngeal carcinoma could be beneficial for the disease.

Chapter 3 - Matrix metalloproteinases (MMPs), also called Martixins, functions in the extracellular environment of the cells and degrade both matrix and non-matrix proteins. They are a large family of zinc-endopeptidases which play vital roles in multiple physiological and pathological processes. Maintenance of the structural integrity of the cardiac extracellular matrix is important for proper functioning of the heart. MMPs were reported to cause changes in the structural framework of the extracellular matrix by stimulation of growth factors and inflammatory mediators, thereby causing abnormality in cardiac remodeling and inflammatory response in cardiovascular diseases. Since most of the MMPs are reported to have their role in pathological shift causing various diseases, they are considered as prominent therapeutic targets for preventive medicine. This chapter brings in the members of MMP family, their structural organization, function and discusses their role in cardiovascular disease.

Chapter 4 - The aim of this paper was to summarize the data about the role of matrix metalloproteinases in the pathogenesis of psoriasis. In psoriasis, matrix metalloproteinases facilitate structural remodeling of the epidermis that, in turn, results in the development of psoriatic plaques. Particularly, they influence the composition of the extracellular matrix and regulate the strength of intracellular contacts between the skin cells. In the dermis, the interaction of matrix metalloproteinases with vascular endothelial cells results in reshaping and vasodilation of venous capillaries. After all, microcapillaries become more permeable for immune cells. At the molecular level, the expression of *MMP1*, *MMP9* and *MMP12* correlates with disease severity. It increases with disease progression and decreases in remission. The separate sections of the paper are dedicated to the role of matrix metalloproteinases in generation of matrikines from the proteins of extracellular matrix and regulation of cytokines that are involved in the pathogenesis of psoriasis. Thus, the authors assessed the contribution of matrix metalloproteinases to the pathogenesis of psoriasis and analyzed changes in physiological processes caused by their differential expression in lesional skin.

Chapter 5 - The degradation of proteins and connective tissue from extracellular matrix can be a crucial event for tissues in intense remodeling process, with proteases as mediators called matrix metalloproteinases (MMPs). Pregnancy is a very peculiar event in which the uterus undergoes intense morphophysiological modifications, especially related to vascular remodeling and angiogenesis, as well as trophoblast invasion, which occurs in early pregnancy period until reaching maternal spiral arteries and replacing the endothelium forming the endovascular-trophoblast. Such vascular modifications ensure the high-flow and low-pressure blood supply to the developing fetus, which in turn requires more and more space, causing the uterus to become enlarged and distended. In this context, an orchestrated regulation of MMPs and their endogenous inhibitors - such as tissue inhibitors of metalloproteinases (TIMP) - are essential for proper gestational development. The disordered uteroplacental remodeling is associated with several obstetric complications, such as placenta accreta, fetal growth restriction, abortion, preterm delivery and pre-eclampsia. Studies of MMPs involved in the uterine remodeling process and their relationship to local maternal immune cells, as well as fetal trophoblastic cells, still represent a very promising field of research, since the elucidation of the involved processes can boost the development of strategies to prevent complications in woman's health, obstetrics and for consequences in offspring adulthood.

Chapter 6 - Bone has a dynamic structure, since it is remodelled during the lifespan to sustain its structure and function. Extracellular matrix (ECM) is playing a tremendously important role, such as cell adhesion, immobilization of growth factors and nucleation of mineralization in bone development phase. It consists of proteins and leads the bone remodelling by the combined osteoblast (bone-forming cells) and osteoclast (bone-resorbing cells) activities. Besides, ECM behaves as a scaffold for mineral deposition. Matrix metalloproteinases (MMPs), a family of zinc-depended proteolytic enzymes, are the most important enzymes used for the degradation of unrelated proteins and structural components present in ECM. MMPs are highly expressed in mammalian bone and cartilage cells and are able to cleave collagens, thus function as collagenases. Furthermore, they lead remodelling of ECM in connection with tissue specific and cell anchored inhibitors. Functions of MMPs may vary bone quality via bone resorption and formation, i.e., osteoblast recruitment and survival, angiogenesis, osteocyte viability and function, chondrocyte proliferation and differentiation. Abnormal expression of MMPs can be related to pathological conditions such as unstable bone remodelling, particularly osteoporosis, rheumatoid arthritis and osteoarthritis. In this chapter, bone tissue components, MMP properties and functions, bone modelling, remodelling and resorption, repair and regeneration, and pathological bone resorption will be discussed.

Chapter 7 - Protein tyrosine phosphatases (PTPs) are important targets that are known to play a key role in the development of chronic degenerative diseases such as obesity, diabetes, and some neurological diseases. To date, different strategies have been developed to produce both reversible and irreversible inhibitors of PTP activity; above all, selectivity has been sought. This has proven to be a challenge for researchers in the field of medicinal chemistry due to the high conservation and cationic nature of the active sites of several PTP. However, small electrophilic molecules with inhibitory activity have been developed to target active sites with high activity and selectivity. Currently, the challenge lies in the design of allosteric and covalent inhibitors to modulate the activity of PTP. The strategy of allosteric and covalent inhibition has generated some successful high-activity small molecules for PTP1B. In this regard, recent advances have led to some specific class of inhibitors. However, none of them have exhibited true selectivity towards a given PTP.

In: A Closer Look at Metalloproteinases
Editor: Lena Goodwin
ISBN: 978-1-53616-517-3

Chapter 1

THE UNFOLDING COMPLEXITY OF THE MATRIX METALLOPROTEINASE (MMP) SYSTEM: RELEVANCE TO DISEASE

A. Hussain[1,*], Y. Lee[2] and J. Marshall[1]

[1]Department of Genetics, UCL Institute of Ophthalmology, London, UK

[2]Alt-Regen Co., Ltd., Heungdeok IT Valley, Yongin, Republic of Korea

ABSTRACT

Extracellular matrices (ECMs) play a pivotal role in regulating the delivery of nutrients and removal of waste products between cells and their blood supply. The structural architecture and composition of ECMs is maintained by tightly coupled processes of continuous synthesis and degradation. Degradation is mediated by a family of Zn^{2+}-containing, Ca^{2+}-dependent proteolytic enzymes called the matrix metalloproteinases (MMPs). These enzymes are released as latent proteins (pro-MMPs) and on activation (following the catalytic removal of a small peptide) are capable of degrading most components of an ECM.

---

* Corresponding Author's E-mail: alyhussain@aol.com.

The catalytic activity of MMPs is modulated by the presence of tissue inhibitors of MMPs (TIMPs) and current opinion holds that the ratio of TIMPs/MMPs determines the relative rate of degradation. Thus, elevated ratios are thought to compromise degradation leading to the accumulation of abnormal ECM material (as in Bruch's membrane of the eye), whilst diminished ratios are thought to lead to excessive ECM degradation allowing the spread of cancer cells, and progression of angiogenesis.

However, this simplistic control system for MMP activity now appears to be far more complex and has implications for our understanding of disease mechanisms involving the MMP system. The MMP species tend to undergo covalent modification leading to homo- and hetero-dimerization and non-covalent aggregation resulting in the formation of very large macromolecular weight MMP complexes (LMMC). In addition, the various MMP species also show a free-bound compartmentalisation. The net result of these changes is to reduce the availability of the latent forms of MMPs for the activation process.

An assessment of the degradation potential of the MMP system in any tissue must therefore take into account the degree of sequestration of the latent MMP species, a protocol that has not previously been addressed. Based on the complexities already described, we will present an analysis of the MMP system in two common neurodegenerative disorders, namely age-related macular degeneration (AMD) and Alzheimer's disease (AD).

Keywords: extracellular matrix, matrix metalloproteinases, protein aggregation, macular degeneration, Alzheimer's disease

INTRODUCTION

Matrix metalloproteinases (MMPs) are a group of Ca^{2+}-dependent, Zn^{2+}-containing proteolytic enzymes that are ubiquitously distributed throughout the body. By regulating the structure and composition of the extracellular matrix (ECM), these enzymes play crucial roles in the homeostatic exchange of nutrients and waste products between cells and their blood supply, and in the modulation of cellular signals that regulate cell proliferation, differentiation, and apoptosis. Thus, their involvement in diverse physiological processes such as embryogenesis, tissue remodelling, and cell growth/migration has been well documented (Pilcher et al., 1999; Chin and Werb, 1997).

Abnormal regulation of MMPs has also been described with pathological outcomes as in inflammation, osteoarthritis, cardiovascular and pulmonary diseases, cancer, etc. (Nakajima et al. 1993; Guan et al., 2003). Recently, attention is being increasingly focused on the role of MMPs in the normal ageing process and in age-related diseases. To better understand the underlying mechanisms in these diseases, we need to delineate the regulatory aspects of the MMP system.

Under physiological conditions, MMPs are released into the ECM by neighbouring cells as inactive or latent pro-enzymes and require the catalytic removal of a small pro-peptide for activation. Once activated, these MMPs are capable of degrading most components of the ECM (Matrisian, 1992; Birkedal-Hansen et al. 1993). The activity of these enzymes is checked by binding of tissue inhibitors of MMPs (TIMPs) to the catalytic site. It should be noted that TIMPs also bind to non-catalytic sites on pro-MMPs 2&9; this binding conferring stability to the enzyme (Howard et al., 1991). Pro-MMPs can also bind to substrates (elastin, gelatine) at non-catalytic sites resulting in partial activation without loss of pro-peptide, and in the case of pro-MMP9, this substrate-mediated activation was 10-fold less than that of the fully activated form (Bannikov et al., 2002).

On a simplistic level, the MMP system may be considered as incorporating pro-MMPs, activated-MMPs and TIMPs with the degradation status being reflected by either the level of activated species or the TIMP/activated-MMP ratio. Since it is difficult to estimate the level of activated MMP species, most studies tend to report the TIMP/pro-MMP ratio, but as indicated below, this does not accurately reflect the MMP degradation potential in the tissue.

Increasing evidence suggests that the regulation of MMP activity is far more complex (Figure 1). Firstly, individual MMPs have now been shown to undergo covalent and non-covalent transformation and secondly, they appear to be partitioned between free and bound compartments. The inter-relationship between the various species and their compartmentation profiles dictates the level of free pro-MMPs available for activation and hence the degradation potential of the system.

A full characterisation of the MMP system requires identification and quantification of individual species and their free/bound distribution. The activation mechanism of MMPs is first described since it is important for identifying the various MMP species. This is then followed by an examination of covalent and non-covalent changes that lead to a variety of MMP species. Next, the inter-relationships between the various MMP species is described culminating in the formulation of the MMP Pathway. Normal ageing changes of the MMP Pathway and their impact on the degradation system is then explored. Finally, the role of the MMP system is assessed in two of the most common age-related diseases, namely age-related macular degeneration (AMD) and Alzheimer's disease (AD).

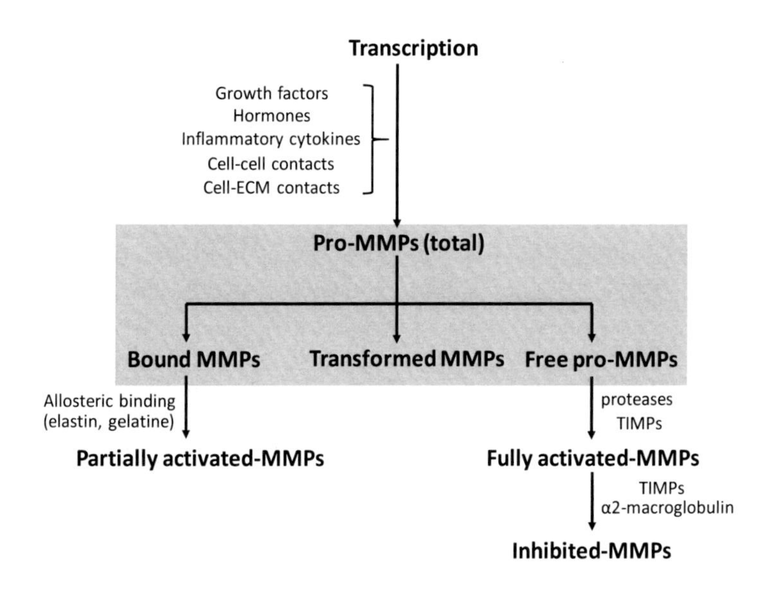

Figure 1. Compartmentalisation of pro-MMPs following release into the ECM. MMP species within the shaded region have been quantified to determine the effect of age on the MMP system and its potential for degradation of the ECM.

ACTIVATION OF PRO-MMPS

MMPs are proteolytic enzymes and their target is the peptide bond, hydrolysis being initiated by nucleophilic attack (Figure 2). Water on its own is a poor nucleophile but coordination with the Zn^{2+} atom within the active site of MMPs makes it a stronger nucleophile and is then able to attack the carbon atom in the peptide bond. This results in bond rearrangements that culminate in the hydrolysis of the protein or peptide.

Pro-MMPs contain a conserved cysteine residue in the N-terminal pro-domain region that coordinates with the Zn^{2+} atom in the catalytic site (-S-Zn^{2+}) preventing the entry of water and at the same time blocking any interaction with the substrate for hydrolysis (Figure 3A). Activation is initiated by a protease that cleaves a small peptide from the N-terminal region of the pro-domain, breaking the -S-Zn^{2+} link, and allowing the entry of water resulting in a partially activated MMP (Figure 3B). The initiation of the activation step is known as the 'cysteine switch' (Springman et al., 1990). There then follows an autocatalytic step that removes the entire pro-domain leading to a fully activated MMP enzyme (Figure 3C). The loss of the pro-domain decreases the molecular weight of the MMP by about 8-12kDa (Park et al., 1991).

The cysteine-Zn^{2+} link can also be broken by organomercurial compounds such as amino phenyl mercuric acetate (APMA) causing the pro-domain to be dislodged leading to autocatalysis and a fully activated MMP enzyme (Nagase & Woessner, 1999).

Allosteric binding of chaotropic agents such as sodium dodecyl sulphate (SDS) cause conformational changes in the pro-MMP that partially 'dislodges' the pro-domain but not enough to activate the enzyme. Now, exchange of SDS with Triton-X100 produces a partially active pro-MMP without any autocatalysis of the pro-domain (Figure 3D) (Woessner, 1995). Thus, the molecular weight of the enzyme does not change but the induced partial activity of the enzyme allows its detection.

Zymography is an electrophoretic technique that allows the detection of MMP enzymes based on their ability to hydrolyse specific substrates. In the case of gelatinases, the substrate incorporated into the gel is gelatine. Electrophoresis is carried out in the presence of SDS containing buffers and thus the SDS-bound proteins are separated according to molecular weight, lower molecular weight proteins running fastest.

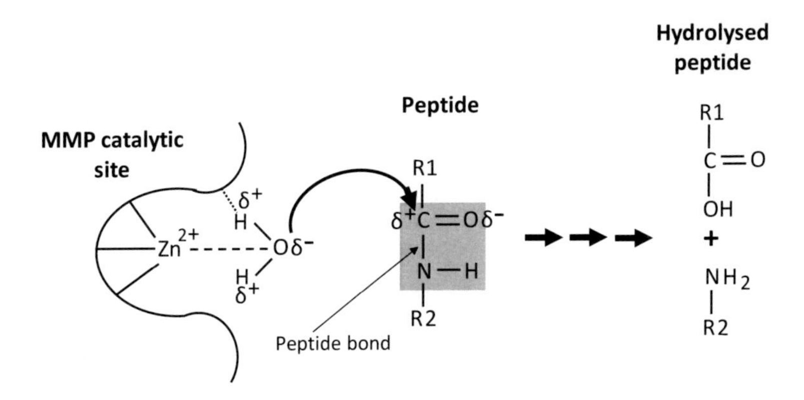

Figure 2. The primary action of activated MMPs. Water in coordination with the Zn^{2+} atom in the catalytic site of MMPs initiates the nucleophilic attack on the carbon atom of the peptide bond leading to the hydrolysis of the peptide.

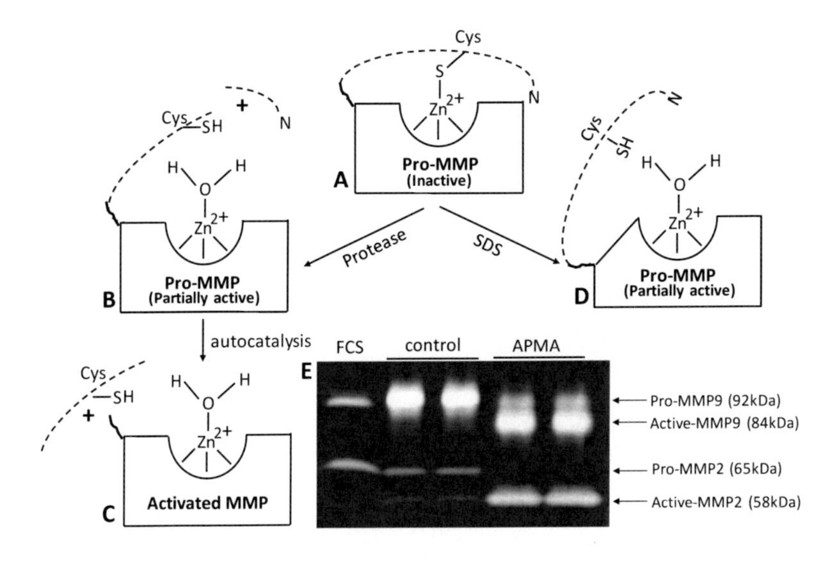

Figure 3. Proteolytic activation of pro-MMPs. In pro-MMPs, the cysteine of the pro-peptide binds the Zn^{2+} atom in the catalytic site, thereby preventing the entry of water or substrate (A). Proteolysis of a small region of the pro-peptide dislodges the remaining pro-peptide allowing entry of water and partial activation (B). This is followed by autocatalysis, removing the entire pro-peptide, leading to full activation (C). Chaotropic agents such as SDS can dislodge the pro-peptide and partial activation occurs without loss of the pro-peptide (D). Organo-mercurial compounds activate pro-MMPs with loss of the pro-peptide resulting in a reduction in the molecular weight of the enzyme (E).

After electrophoresis, the gel is thoroughly washed with Triton X-100 to allow removal of SDS and to 'renature' the proteins. Incubation of the gel in Ca^{2+} containing buffers for a period of 18-40 hours at 37°C allows the gelatine within the enzyme bands to be hydrolysed. Because the samples were prepared and electrophoresis undertaken in SDS buffers, pro-MMPs will be partially activated (as explained above) allowing their detection on zymographic gels. After incubation, the gel is stained with Coomassie Blue whereby the unhydrolyzed gelatine containing regions are stained dark blue and the hydrolysed regions (due to MMP activity) remain colourless (Figure 3E). Some investigators prefer to invert the colours of the photographed gel so that MMP bands now appear dark against a whitish background.

For electrophoresis, samples are prepared in SDS containing buffers. Therefore, complexes that are held together by hydrophobic or electrostatic interactions will be disassociated and other methods are required for their detection. Thus, TIMP inhibitors that may be bound to MMPs will be separated and because of their much lower molecular weights, will run much faster in the gel. These can also be detected by the technique of 'reverse' zymography.

MODEL SYSTEM FOR STUDYING MMP REGULATION

We would require a tissue that is readily available, maintains a high degradative capacity, is amenable to biophysical manipulation so as to easily assess functional parameters, and obtainable at various stages of the ageing process. Bruch's membrane of the eye fulfils these criteria. It is a penta-laminated ECM structure, lying between the blood supply in the choriocapillaris and the retinal pigment epithelium (RPE)-photoreceptor complex (Figure 4).

The major MMP species present in Bruch's are the gelatinases MMP2 and MMP9, the former being the constitutive enzyme and the latter the inducible form (Guo et al., 1999). These MMPs enter Bruch's from primarily the RPE with a small contribution from choroidal cells. The membrane can be removed intact as a flat preparation, free from contaminating MMP-producing cells. Such a preparation can be perfused to obtain free and bound MMP fractions. Furthermore, the free fractions can be fractionated on a gel-filtration column according to size (without disrupting hydrophobic/electrostatic interactions) to identify the various non-covalently bonded MMP species. Obtaining Bruch's from donors of different ages also allows the examination of the effect of age on the MMP system.

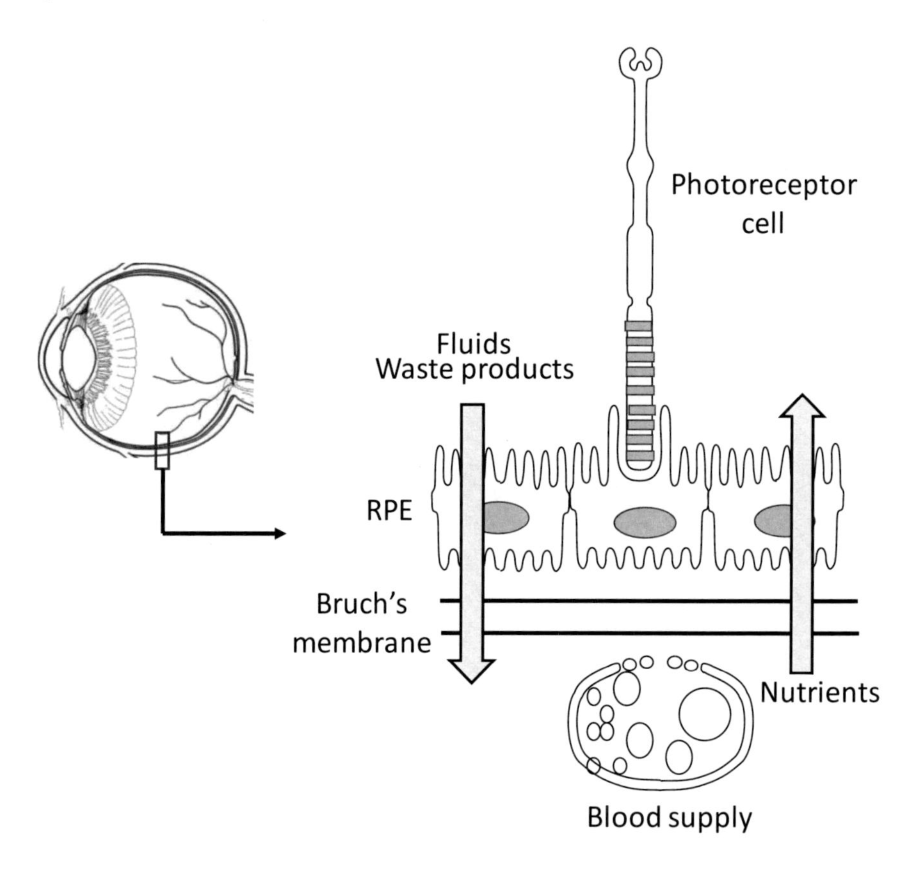

Figure 4. Schematic of the visual unit of the eye comprising photoreceptor cells, retinal pigment epithelium (RPE), Bruch's membrane, and the blood supply. Photoreceptor survival in dependent on efficient bi-directional exchange of nutrients and waste products across Bruch's requiring an efficient MMP system to maintain the integrity of the membrane; abnormalities can lead to photoreceptor death and blindness.

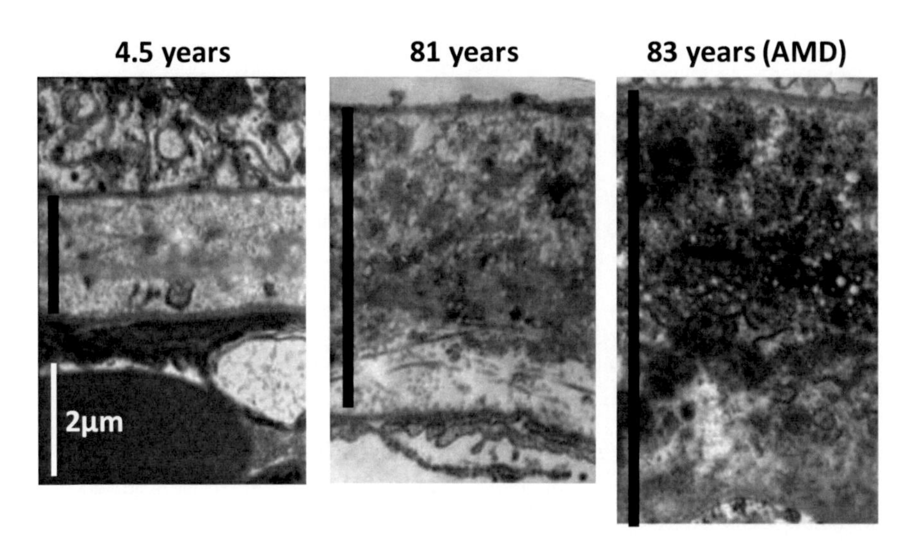

Figure 5. Electron micrographs to show changes in the thickness of Bruch's membrane in normal ageing and AMD. Thickness of the membrane is shown by the black vertical bars. Ageing results in increased thickness from ~ 2.0µm in the young to about 5.5µm in the elderly with much advanced changes in AMD.

Another important reason for studying Bruch's membrane is the role it plays in normal physiology and pathology associated with age-related macular degeneration, a condition that leads to blindness. Bruch's membrane houses the transport pathways for the bidirectional movement of nutrients and waste products and its functional status is therefore important for the survival of photoreceptor cells. Waste products include membraneous debris and toxic lipid metabolites that serve to cross-link and 'clog' the membrane. Ageing is associated with increased thickness of the membrane and deposition of lipid rich proteinaceous debris that is expected to interfere with the functional aspects of Bruch's (Figure 5).

As such, the functional parameters of Bruch's membrane decline exponentially with age with a half-life of ~ 18 years, i.e., the capacity to transport fluids and carrier-laden proteins is halved for every 18 years of life (Figure 6). In the elderly, these decay curves approach the failure threshold leading to problems associated with low delivery of vitamin A. Crossing the thresholds (as in AMD) leads to degenerative changes.

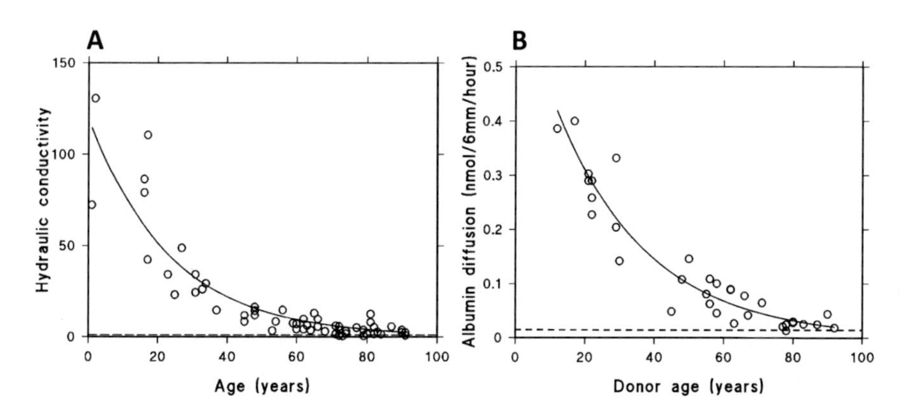

Figure 6. Age-related changes in the hydraulic conductivity (A) and diffusion of albumin (B) of human Bruch's membrane. Both transport phenomena showed an exponential decline with age (After Hussain et al., 2004; Lee et al., 2015). The dashed horizontal line represents failure threshold.

Since the MMP system is responsible for rejuvenation of Bruch's membrane, the age-related deposition of debris and exponential decline in transport functions implies problems with the degradation process. It is therefore important to assess the MMP system in both normal ageing and AMD.

Covalent Modification of MMP Species

The gelatinase species present in Bruch's membrane comprise MMPs 2 & 9 and have previously been identified by immunochemical methods (Guo et al. 1999). Since each of these species can give rise to the activated form, we would expect to observe a total of four gelatinase species on zymograms. However, zymographic analysis of extracts from Bruch's membrane show the presence of six gelatinase species, the additional two of higher molecular weight being termed HMW1 and HMW2, corresponding to molecular weights of 122±9 and 344 ± 22kDa respectively (Figure 7, Mean ± SD, n = 24 donors) (Hussain et al., 2010a; Kumar et al., 2010). These high molecular weight species have previously been observed in body fluids and a variety of tissues including brain (Backstrom et al., 1992; Lim et al. 1997; Roomi MW et al. 2015; Yan et al., 2001).

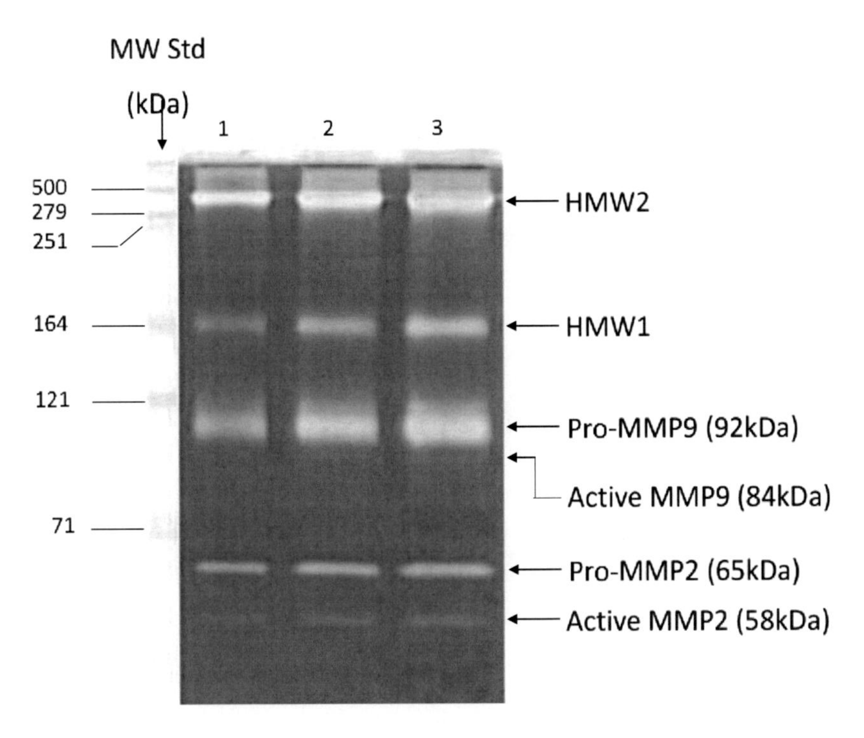

Figure 7. Gelatinase species in extracts of Bruch's membrane obtained from two donors aged 75 and 83 years. Altogether, six gelatinase species have been identified, high molecular weight species HMW1 & HMW2, and pro- and active-forms of MMPs 2&9. Lanes 1,2,3 are incremental loading of sample (From Hussain et al., 2010a).

In non-ocular tissues, HMW1 has been shown to be a complex of pro-MMP9 and neutrophil gelatinase-associated lipocalin (NGAL) (Triebel et al., 1992). Incubation of pro-MMP9 with NGAL resulted in the production of the pro-MMP9-NGAL complex (Yan et al., 2001). The lipocalin family includes α1 and α2-microglobulins, α1-acid glycoprotein, complement C8γ, and retinol binding protein (Flower, 1996). The identity of the lipocalin that covalently binds to pro-MMP9 in Bruch's membrane is not known but retinol binding protein which occurs in high concentrations in Bruch's membrane is a good candidate. Although complexed with lipocalin, the inhibitory pro-domain of pro-MMP9 can be removed by APMA, leading to activation and a slight reduction in molecular weight. Whether HMW1 can function as a gelatinase *in vivo* has yet to be determined.

In Bruch's membrane, fragmentation studies of HMW2 have shown it to consist of covalently bonded pro-MMPs 2&9 although the stoichiometry in the complex is not known (Hussain et al. 2010a). A similar high molecular weight MMP species was observed in brain tissue and on APMA activation, its molecular weight was reduced by 10kDa indicative of the removal of the pro-domain peptide from one of the MMP molecules in the complex (Lim et al., 1997). Again, it is not known whether this 'activated' complex can function as a gelatinase *in vivo*.

Disulphide-bonded dimers of pro-MMP9 (MW 215-220kDa) have been identified in both normal and tumour cells (Triebel et al., 1992; Upadhya et al.. 1997; Kjeldsen et al. 1993) but not observed in ocular or brain tissue (Hussain et al., 2017)). Their increased secretion in carcinomas, sarcomas, and leukemia cell lines has suggested a possible role for this dimer in cellular migration (Roomi et al. 2014; 2015; Dufour et al., 2010). The formation of the covalently bonded high molecular weight species serves to effectively remove free pro-MMPs from the activation step and is therefore expected to have implications on the degradation potential of these MMPs.

NON-COVALENT AGGREGATION OF MMP SPECIES

The non-covalent interaction between TIMPs and MMPs 2 & 9 occurs at two locations on the enzyme, at the catalytic site and at the carboxy terminal of the hemopexin-like domain, away from the catalytic site (Kleiner et al., 1993; Willenbrock et al., 1993). In the presence of SDS, the MMP-TIMP complex is disassociated because of the non-covalent attachment (Woessner, 1995). Thus, on zymographic gels, loosely held aggregates (due to hydrophobic or electrostatic interactions) such as the MMP-TIMP complexes or dimeric or multimeric pro-MMP species will be dispersed giving rise to monomeric species that can be characterized by their respective molecular weights.

Analysis of loosely held complexes requires the use of less harsh techniques such as gel filtration chromatography that maintain the native configuration of the complex.

In gel-filtration chromatography, a suitable separating medium such as Sepharose CL-6B in the form of beads of diameter 40-165μm is packed into a glass column and equilibrated with a physiological buffer. Such a column can fractionate proteins or protein complexes in the molecular weight range 10kDa to 4000kDa, the bigger species eluting first followed by the smaller species. Protein complexes that are larger than 4000kDa cannot enter the beaded material for fractionation but simply pass in-between the beads and are eluted very early at the forefront of the chromatography run. The procedure is shown schematically in Figure 8.

In the chromatographic run, samples isolated in physiological buffers (thereby maintaining the integrity of loosely-held complexes) are loaded onto the column and eluted with the same physiological buffer (Figure 8A). A certain amount of fluid (called the void volume, Vo) has to exit the column before the very large complexes (that could not enter the gel beads) leave the column. The elution profile is expressed as a ratio (Ve/Vo) where Ve is the cumulative elution volume. Thus, a Ve/Vo ratio of 1 signifies the elution front where unfractionated large complexes appear. This is followed by the release of fractionated proteins, the larger exiting first (Figure 8B). The collected fractions are then subjected to gelatine zymography to identify the monomeric MMP species present (Figure 8C).

A densitometric scan is then undertaken for each MMP species and plotted against the Ve/Vo ratio (Figure 8D). The peak Ve/Vo of a given species is used to determine the molecular weight of the species with reference to a calibration plot. A comparison of the molecular weights between the zymographic (monomer) and gel filtration methods (possible aggregates) allows the designation of a given MMP species, present in the ECM, as monomeric, dimeric, or multimeric.

This methodology requires a lot of tissue if the MMPs are to be detected in the much diluted eluants by zymography.

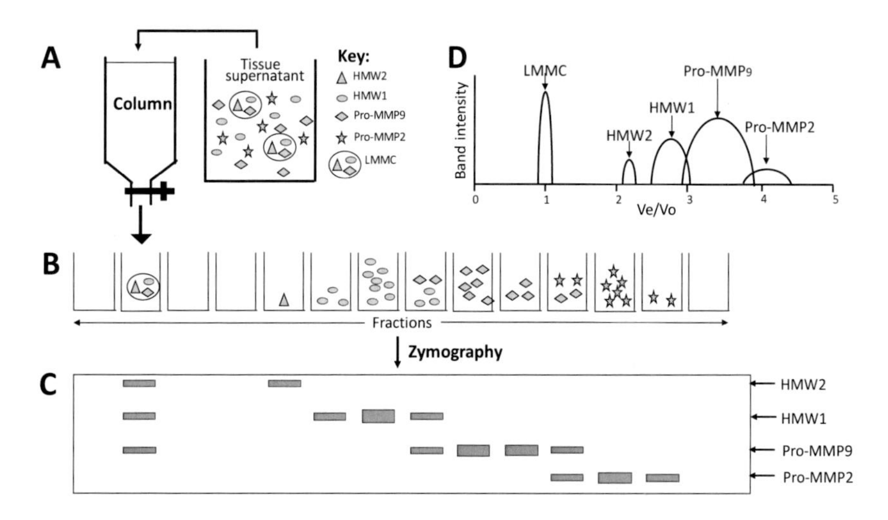

Figure 8. Schematic to show the fractionation of MMP species retaining their native configuration using gel filtration chromatography. A sample of tissue supernatant is loaded onto the column (A) and elution results in the separation of MMP species according to their molecular weight (B). The fractions (10-50, 1.0ml each) are then subjected to zymography to identify the monomers present in each complex (C). A plot of gel band intensity versus Ve/Vo shows the relative elution of these species, allowing the calculation of their molecular weights (D).

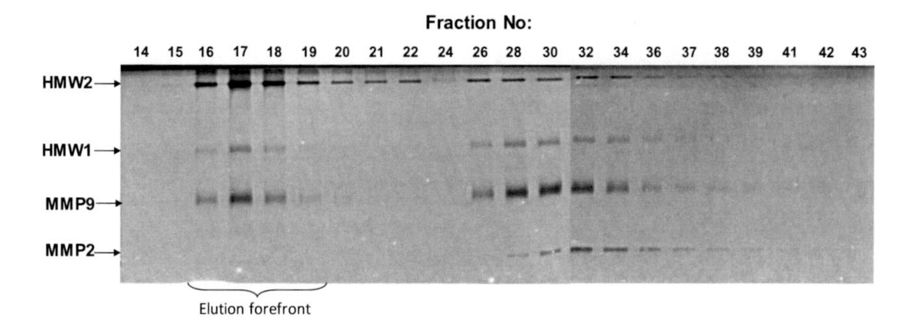

Figure 9. Gel-filtration and zymographic analysis of MMP entities in the supernatant fraction of human Bruch's membrane. At the elution forefront (fractions 16-19), HMW2, HMW1, pro-MMP9, and a trace of pro-MMP2 were present, and must have travelled as a complex (LMMC) of MW > 4000kDa. Free MMP species were then eluted according to their molecular weights. (From Hussain et al., 2010a).

For a representative analysis, Bruch's membrane from eight donors (age 69-84 years) was pooled, homogenised, and the soluble supernatant fraction subjected to gel filtration chromatography. The eluants were then analysed by zymography using five gels, and the three gels showing the presence of MMP species have been collated and shown in Figure 9.

In Figure 9, fractions 16-19 represent a Ve/Vo ratio of ~ 1.0, i.e., these eluants should contain molecular structures that were too large to be fractionated on the column. The presence of HMW2, HMW1, pro-MMP9 and occasionally pro-MMP2 in this region implies that these MMP species must have travelled as a combined complex (see Figure 8). This combination of MMP species has been termed the Large Macro-Molecular weight MMP Complex (LMMC). Its presence at the forefront of the chromatographic run means that its molecular weight must exceed 4000kDa. Its release on homogenisation of Bruch's membrane suggests that the complex was not bound to the ECM but may have been trapped within the fibrillar matrix. Also note that in fractions 20-36 there is some presence of HMW2 and this may be related to a spectrum of LMMC particles of diminishing molecular weight.

Release of the LMMC complex was followed by the appearance of free MMP species and a plot of densitometric intensity of the bands versus Ve/Vo was used to obtain the molecular weight of individual species with reference to a calibration curve. For a statistical comparison of molecular weights of a given MMP species using gel filtration chromatography (native configuration) and zymography (monomer), the above analysis of Figure 9 was repeated using preparations from 6-8 donors, age range 86-95 years. The ratio of molecular weights determined by gel filtration (GF) and zymography (Z) was calculated for each species; a ratio of ~ 1.0 signifying a monomer, ~ 2.0 signifying dimer, and > 3.0 signifying a multimer (Table 1). Pro-MMP2 and HMW2 exhibited similar molecular weights by zymography and gel filtration, and were therefore considered to exist as monomers in the ECM. On the other hand, pro-MMP9 and HMW1 were shown to exist as loosely bound dimers.

The data shows that the soluble compartment of Bruch's membrane contains highly complexed MMP multimers (LMMC), dimers of pro-MMP9 and HMW1, and monomers of pro-MMP2 and HMW2.

EXISTENCE OF A BOUND-FREE COMPARTMENT OF MMPS

The previous section has shown that MMPs are complexed forming the LMMC particle. It is also likely that individual MMP species are also bound to the matrix or trapped within the accumulated debris within Bruch's membrane. The likelihood of such a free-bound distribution can be evaluated using perfusion experiments. Perfusion is expected to release the free pool and subsequent analysis of the remaining tissue should be indicative of the bound pool.

Table 1. Native configuration of gelatinase species in Bruch's membrane

	Zymography (Z)	Gel filtration (GF)	Ratio (GF/Z)	State in tissue
Pro-MMP2	65 ± 1kDa	88 ± 14kDa	1.3	Monomer
Pro-MMP9	95 ± 2kDa	191 ± 35kDa	2.0	Dimer
HMW1	122 ± 9kDa	281 ± 12kDa	2.3	Dimer
HMW2	344 ± 22kDa	307 ± 17kDa	0.9	Monomer

A Bruch's preparation is mounted in an open-type Ussing chamber and perfused at intra-ocular pressure with eluant samples being collected every hour for a period of 5-10 hours (Figure 10A). The time for cessation of MMP release is dependent on the hydraulic conductivity of the membrane. Bruch's from older donors has low hydraulic conductivity and therefore it takes much longer to release the free pool (see Figure 6).

The eluted samples are subjected to zymography to identify and quantify the released MMP species. Since the LMMC complex is far too large to be released (> 4000kDa), its contribution to the free pool is expected to be minimal. Perfusion resulted in the appearance of most MMP species in the eluant samples (exception being active-MMP9) and the amount released diminished over time until none were detected after 5-10 hours of perfusion (Figure 10B). At the end of the perfusion period, a surgical trephine was used to remove the perfused membrane (containing bound MMP species) and zymographic analysis showed the copious presence of all MMP species demonstrating the presence of a free-bound equilibrium (Kumar et al., 2010).

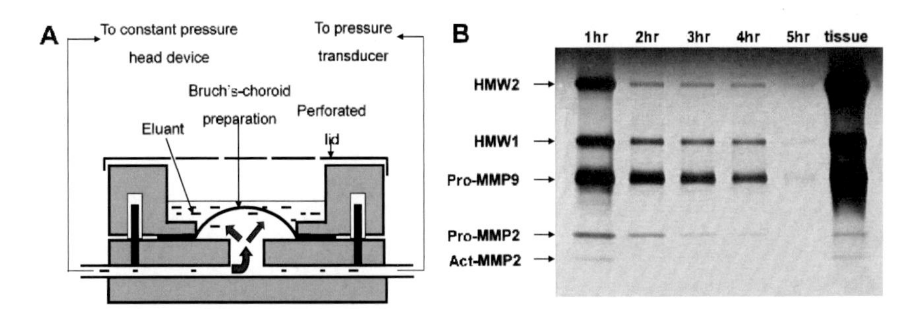

Figure 10. Perfusion of Bruch's membrane to separate free and bound MMP species. An isolated Bruch's membrane preparation is clamped in an Ussing chamber and perfused with phosphate buffered saline at intra-ocular pressure (A). Fluid traversing the membrane is collected every hour and together with the tissue sample obtained at the end of the experiment examined for presence of MMPs by zymography (B).

Effect of Ageing on the Free-Bound Compartment of MMPs

Perfusion experiments are quite cumbersome to carry out and often require long periods to complete. An alternative method for determining bound and free levels of MMP species is to homogenise or disintegrate the tissue and then to examine supernatant (free) and pellet (bound) fractions. Unlike perfusion, this method allows the examination of a large number of samples, allowing the construction of age profiles. However, homogenisation will also release the LMMC complex but its contribution to the free pool can be taken into account in the following manner.

Gel filtration columns were previously used to separate LMMC from the free MMP pool and therefore, the percentage of total MMPs in the supernatant fraction that were resident in the LMMC portion could be calculated. The LMMC complex from Bruch's was observed to contain the following percentages of MMP species present in the supernatant: HMW1 (56 ± 34%), HMW2 (89 ± 13%), pro-MMP9 (33 ± 8%), with little or no pro-MMP2 (Hussain et al., 2010a). Thus, virtually all HMW2 and about 50% HMW1 present in the supernatant resides within the LMMC complex, whereas most of pro-MMP9 and all of pro-MMP2 remain in free solution.

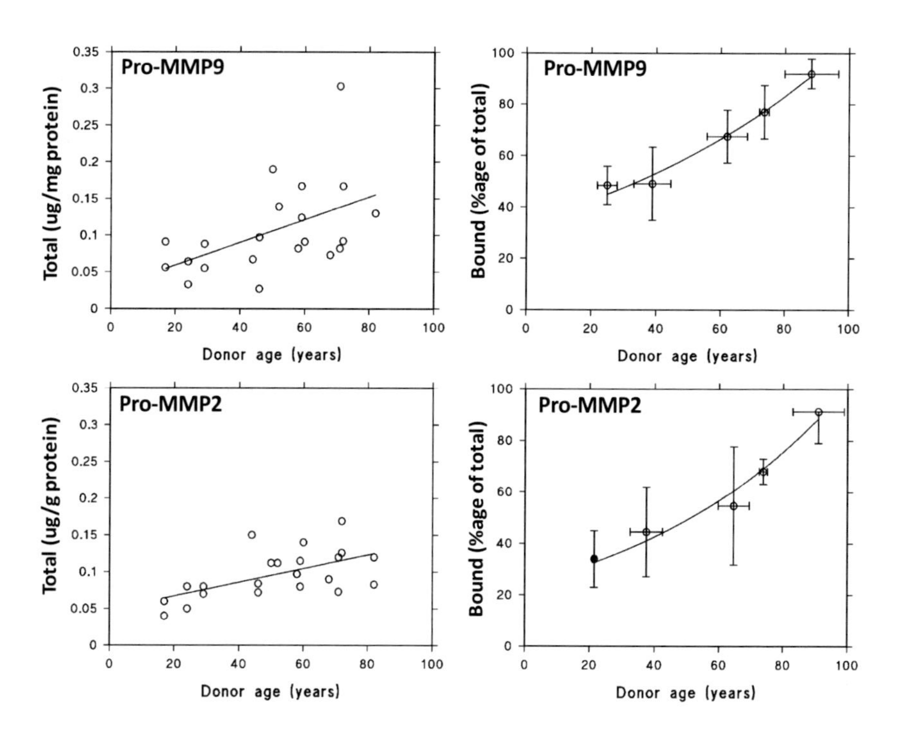

Figure 11. Age-related variation of pro-MMP2&9 in total and bound compartments of Bruch's membrane. Ageing was associated with a linear increase in total levels but an exponential increase in the amount bound. (22 donors, age range 18-82) (Re-calculated from Guo et al., 1999; Kumar et al., 2010; Hussain et al., 2010a).

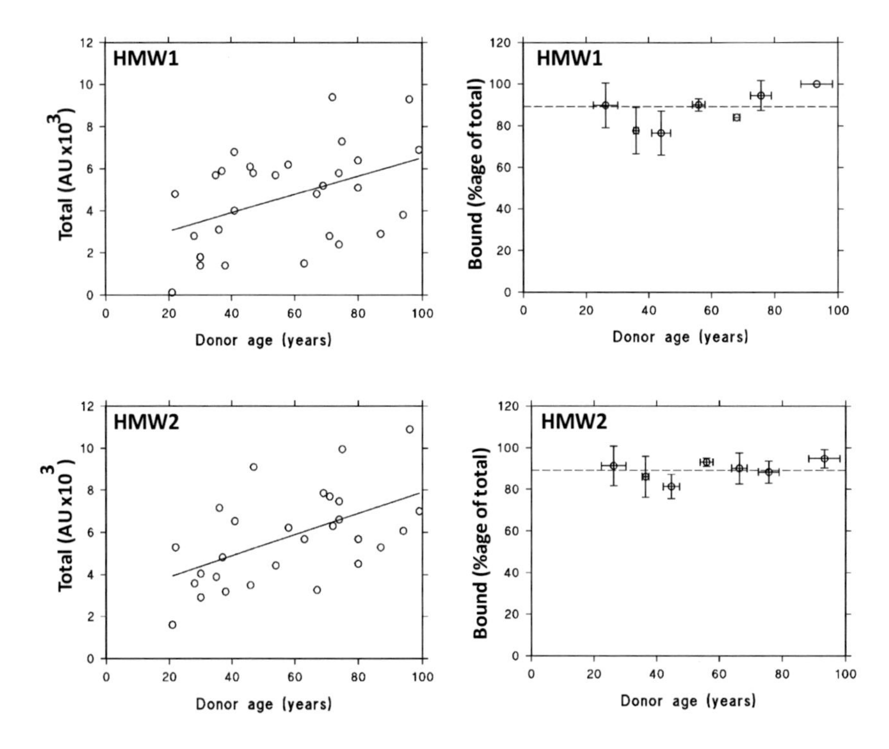

Figure 12. Age-related variation of HMW1 & HMW2 in total and bound compartments of Bruch's membrane. Ageing was associated with a linear increase in total levels but the percentage bound remained constant (~ 90%) (29 donors, age range 21-99). AU = arbitrary units. (Re-calculated from Kumar et al., 2010 and Hussain et al., 2010a).

The total amount of pro-MMPs 2&9 in Bruch's membrane was observed to increase linearly with age. However, the amount bound to the extracellular matrix was observed to increase in an exponential manner, thereby diminishing the free pool of pro-MMPs, and reducing the potential for activation (Figure 11; Guo et al., 1999; Kumar et al., 2010; Hussain et al., 2010a). Bound levels of pro-MMP9 increased from 42% of total at 20 years of age to 83% at 80 years. Similarly, bound levels of pro-MMP2 increased from 34% to 75% between the ages of 20 and 80 years.

Pro-MMP9 undergoes covalent complexation with lipocalins and pro-MMP2 to yield HMW1 and HMW2 respectively. The increase in the content of pro-MMP 9 in Bruch's with age is expected to also affect the level of the high molecular weight species. Thus, in Bruch's, the total level of HMW1 and HMW2 were observed to increase linearly with age (Figure 12). Interestingly, the percentage of bound HMW1 & HMW2 remained invariant with age and was about 90%. Since most of these species present in the supernatant fraction were bound within the LMMC particle, the results show that virtually all HMW1 and HMW2 of Bruch's membrane resides in the bound compartment.

The MMP Pathway

The gelatinase species within Bruch's membrane therefore includes pro-MMPs 2&9, their activated counter parts, MMPs 2&9, high molecular weight components, HMW1 and HMW2, and the LMMC complex. The inter-relationships between these species is represented as the MMP Pathway (Figure 13; Hussain et al., 2011; Hussain et al., 2014, Hussain et al, 2017). Apart from the LMMC complex, pro-MMPs 2&9, and HMW 1&2 also exist in a free-bound equilibrium.

The MMP Pathway is dynamic with the primary driving force being age. With ageing, the MMP Pathway is shifted to the left utilising free pro-MMPs to generate the higher molecular weight species and thereby diminishing the free pool for activation. In this scenario therefore, ageing is expected to reduce the MMP degradation potential leading to abnormalities in the turnover and function of the ECM.

Having established the MMP pathway, it is now possible to examine the underlying mechanisms involved in regulating the movement of MMP species in the pathway as a function of ageing. This is best illustrated with reference to the activation of pro-MMP2, the predominant gelatinase for matrix turnover.

Pro-MMP2 is activated on the basolateral surfaces of the retinal pigment epithelium (RPE). A trans-membrane MMP called MMP14 initially binds TIMP2 to form a binary complex, and this promotes the binding of free pro-MMP2 to form a ternary complex. A second molecule of MMP14 then cleaves the pro-domain of pro-MMP2 to release the activated-MMP2

(Figure 14) (Strongin et al., 1995; Smine & Planter, 1997; Butler et al., 1998). Therefore, in this cycle, the driving force for activation is the free level of pro-MMP2 and TIMP2. As shown in Figure 6, the transport properties of Bruch's membrane decline exponentially with age, and therefore the diffusional status of both pro-MMP2 and TIMP2 will be reduced, leading to lower activation of the enzyme.

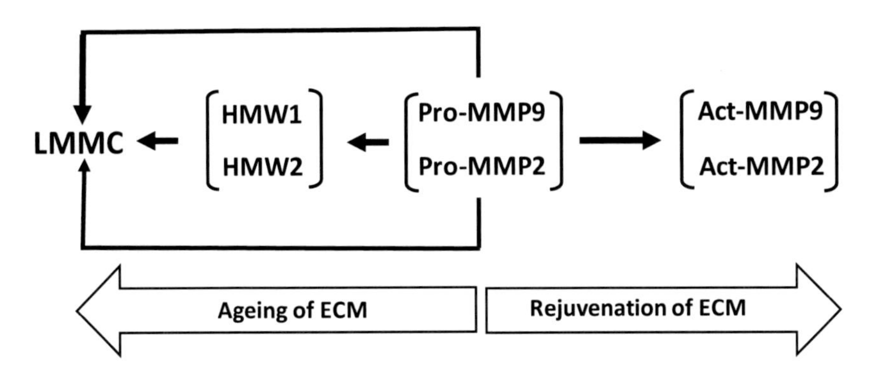

Figure 13. The MMP Pathway in Bruch's membrane. A shift of the pathway to the left is associated with ageing of the ECM leading to structural and functional abnormalities. A shift of the pathway towards the right is required to rejuvenate Bruch's membrane.

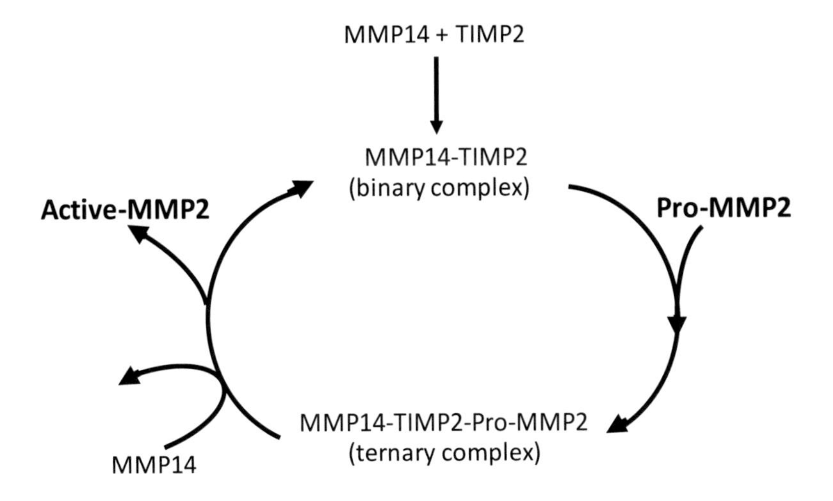

Figure 14. Mechanism for activation of pro-MMP2. The driving forces are the free levels of TIMP2 and pro-MMP2.

In addition to the requirement of sufficient levels of TIMP2 and pro-MMP2 for activation, other processes that compete for pro-MMP2 such as incorporation into LMMC, covalent modification to form HMW2, and binding to the ECM will also influence the degree of activation of the enzyme (Figure 15).

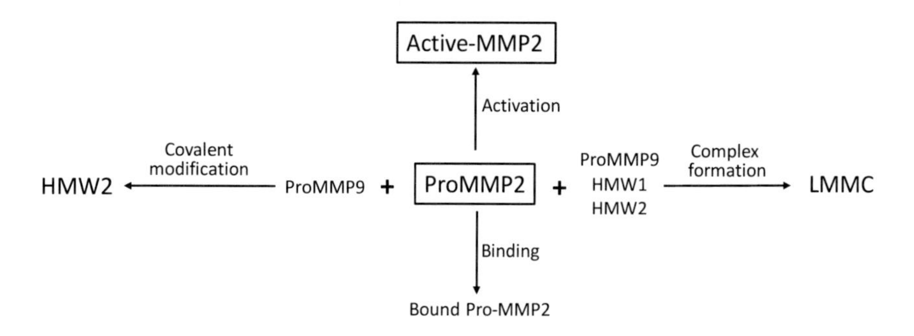

Figure 15. Competitive mechanisms that serve to deplete the free level of pro-MMP2, diminishing the potential for activation.

The MMP System in Age-Related Macular Degeneration

Bruch's membrane plays a crucial role in photoreceptor maintenance by housing transport pathways that mediate the bi-directional delivery of essential nutrients and highly toxic waste products between the choroidal blood supply and the photoreceptor/RPE complex (Figure 4). Ageing is associated with increased thickness of the membrane (Ramratten et al., 1994), large deposition of lipid material (Holz et al., 1994), accumulation of toxic retinoid and lipid oxidation products, increased cross-link formation on collagen (Karwatowski et al., 1995), and increased presence of advanced glycation and lipid end-products (AGEs & ALES) (Handa et al., 1999).

In the normal elderly, the amount of denatured and abnormal collagen resident within Bruch's membrane accounted for nearly 50% of total collagen (Karwatowski et al., 1995). The ageing changes were associated with an exponential decline in both fluid transport and diffusional exchange of nutrients and toxic products across the membrane (Figure 6) (Starita et al., 1996; Hussain et al., 2002; Hussain et al., 2010b).

Despite this large reduction in transportation across Bruch's, most elderly subjects do not present with signs of obvious visual disturbance. However, elderly subjects do show delayed recovery of retinal sensitivity after a bright flash, thought to be due to insufficient levels of retinoids in the RPE for transport to the photoreceptor cells (Patryas et al., 2013; Jackson et al., 1999). These levels in the RPE are normally maintained by efficient transport of vitamin A across Bruch's, a transport process that is severely curtailed with ageing of donor. The delays in dark adaptation after a bright flash can be corrected by administering mega-doses of vitamin A, so as to temporarily overcome the transport barrier in Bruch's (Owsley et al., 2006; Jacobson et al., 1995).

In the normal ageing population, the considerable elevation in the amount of denatured collagen deposited within Bruch's suggests problems with the MMP degradation system. As previously explained, the MMP pathway is shifted to the left on ageing, leading to the accumulation of high molecular weight MMP species and a concomitant reduction in the free pool of MMPs 2&9 for the activation stage. The net result is a reduction in active-MMPs for matrix degradation (Guo et al., 1999).

Other possibilities for reduced matrix degradation have also been suggested. Thus, the accumulation of AGEs and ALEs, known to be potent inhibitors of active MMPs (Nagai et al., 2009; Mott et al., 1997), increase in inter-molecular cross-links known to reduce the susceptibility of the collagen molecule to proteolytic action (Hamlin and Kohn, 1971; Vater et al., 1979), and the age-related increase in levels of TIMP3 (Kamei and Hollyfield, 1999) (inhibitor of active-MMPs) may all contribute to the observed reduction in matrix turnover. If this was the case, then the exogenous addition of activated MMPs to a preparation of isolated Bruch's membrane should be without effect. But we have shown that such an intervention leads to rapid matrix degradation with considerably improved transport properties of Bruch's membrane (Ahir et al., 2002).

In normal ageing therefore, the problem appears to be insufficient levels of activated MMPs and/or entrapment of these species within the grossly altered collagenous network.

In advanced ageing associated with AMD, the structural and functional alterations of Bruch's membrane are much exaggerated (Figure 5), leading to a further decrease in transport capability. Because of the gross structural alterations in the macular region of AMD donors, it has not been possible to isolate intact preparations of Bruch's for transport studies. However, peripheral samples have been studied (Figure 16). The diffusion of a 21kDa FITC-dextran probe across Bruch's was considerably reduced and values for all four AMD donors examined were outside the 95% confidence limits of the control population. In comparison to macular regions, peripheral regions of AMD patients show very little evidence of structural or functional abnormalities. Thus, the reduction in diffusion in peripheral regions suggests that corresponding changes at the macula must be severe.

These severe alterations impact on the survival of the visual system in the following manner. Firstly, the compromised nutritional delivery is expected to provide an initial metabolic insult to the RPE. The delivery components include essential fatty acids, antioxidants, and divalent metals that participate in the anti-oxidant machinery of both the RPE and photoreceptor cells. A curtailment in this supply will increase the oxidative stress leading to greater production of toxic metabolites. Secondly, the stagnation of transport through Bruch's will increase the level of pro-inflammatory and neovascular mediators leading to the death of the RPE and photoreceptor cells, culminating eventually in blindness.

The MMP system is central in the degradation of normal and abnormal components of Bruch's membrane and its potential role in the degenerative changes in AMD have been examined. Because of the severe limitation in the amount of tissue available, it has not been possible to assess the level of the LMMC complex. Nonetheless, data from gel-filtration studies with normal tissue can be used to estimate the level of LMMC in AMD samples as follows. We know that nearly 90% of the HMW2 species exists bound to the matrix, the remaining 10% is sequestered by the LMMC complex (see earlier). If in AMD samples, we observe supernatant levels of HMW2 to be greater than 10% of total, we have to assume that this represents an increase in the level of the LMMC complex.

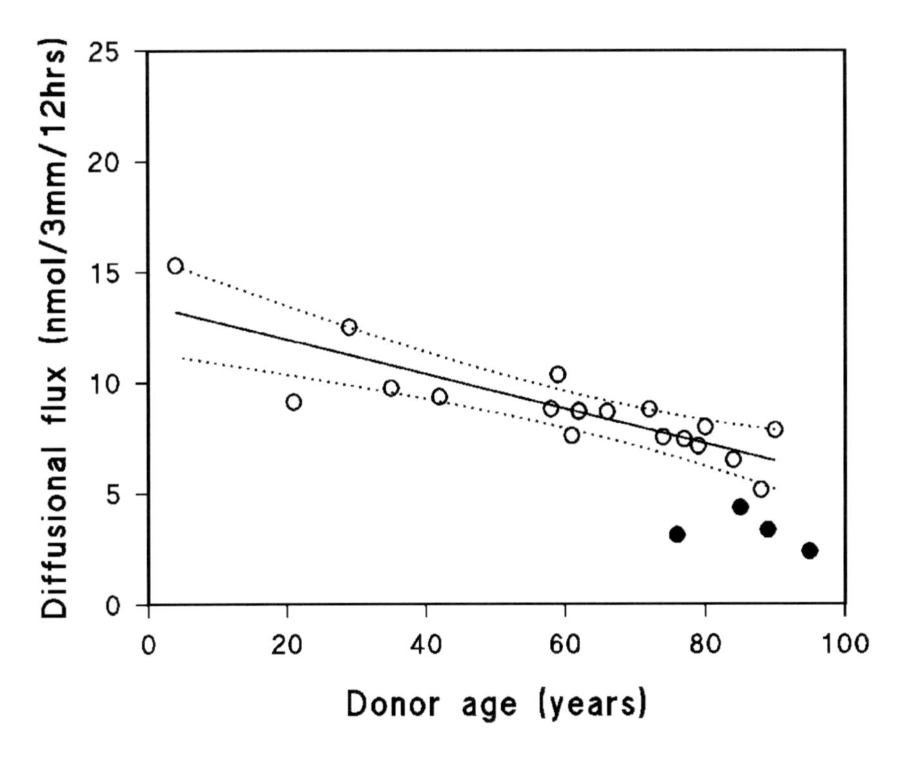

Figure 16. The age-dependent diffusion of a 21kDa FITC-dextran probe across peripheral Bruch's membrane in normal and AMD donors. Open circles, normal donors; filled circles, AMD donors (Hussain et al., 2010b).

The total amount (bound + free) of HMW1 and HMW2 species was increased in Bruch's membrane from AMD donors ($p < 0.05$) indicating a shift of the MMP Pathway to the left compared to age-matched controls (Figure 17; Hussain et al., 2011).

Free or supernatant levels of HMW2 were also raised, representing 31% of total ($p < 0.05$). Since we would expect the supernatant level of HMW2 to be about 10% of total, the higher percentage observed in AMD signifies increased levels of the LMMC complex. Also increased was the level of pro-MMP9, particularly in the supernatant or free fraction ($p < 0.005$). Since ~ 30% of this can be attributed to the increase in LMMC, the remainder, truly free form was elevated at least 2.5-fold compared to controls (Figure 17). Thus, with reference to Figure 15, elevations in free pro-MMP9 would promote the greater formation of HMW2, reducing the free level of pro-MMP2 available for activation. The reasons for elevated levels of pro-MMP9 are not known but may be related to the more frequent presence of a polymorphism in the promotor region of the MMP9 gene in AMD patients (Ye, 2000; Shimarjiri, et al., 1999). The presence of this polymorphism is associated with higher expression of pro-MMP9 in both plasma and Bruch's membrane (Fiotti et al., 2007; Chau et al., 2008; Hussain et al., 2011).

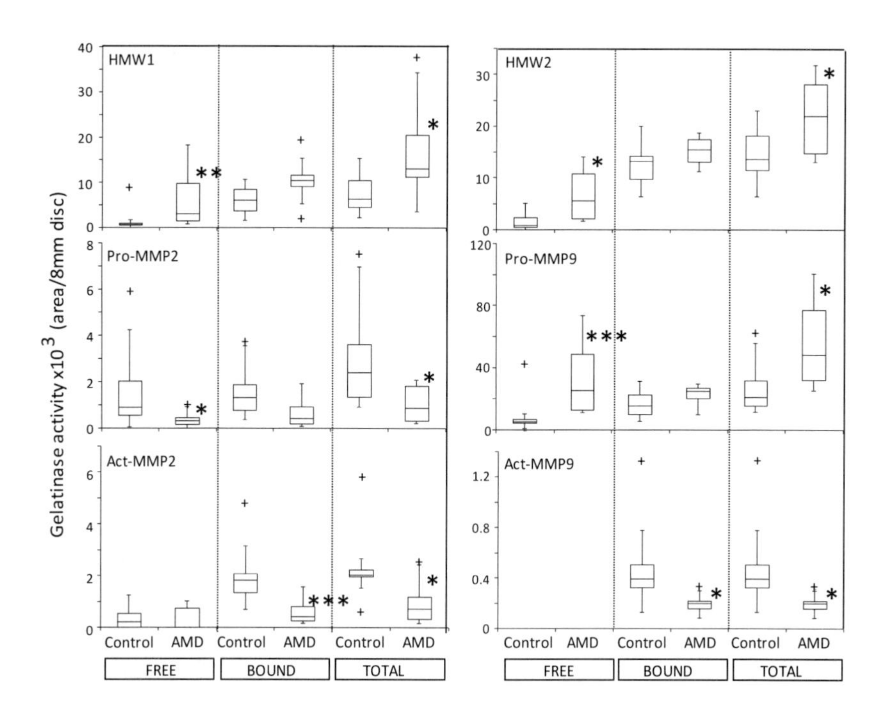

Figure 17. The relative distribution of free and bound MMP species of Bruch's membrane from age-matched control and AMD donors. The free level implies the level in supernatants and thus includes the contribution from the LMMC complex. Data obtained from 13 control (age-range 71-99 years) and 6 AMD donors (age-range 71-95 years). Statistical analysis by the Mann-Whitney non-parametric test. * $p < 0.05$; ** $p < 0.01$, *** $p < 0.005$. (From Hussain et al., 2011).

THE MMP SYSTEM IN ALZHEIMER'S DISEASE (AD)

The extracellular deposits present in AMD and AD share similar characteristics. Common constituents within the deposits include vitronectin, apolipoprotein E, amyloid P, amyloid β (Aβ), lipids, inflammatory mediators, and complement components (Crabb et al., 2002; Anderson et al., 2004; Isas et al., 2010). Whereas AD shows the abundant presence of aggregated Aβ peptides in the ECM, AMD is characterised by the deposition of lipids, lipoproteinaceous debris, and oxidised and damaged ECM components within Bruch's membrane (Karwatowski et al., 1995; Curcio et al., 2005).

Aβ is synthesised by similar pathways in both brain and the RPE involving the sequential cleavage of the membrane bound amyloid beta precursor protein (AβPP) by β-secretase (ACE1) and the presenilin-dependent γ-secretase. Proteolysis by γ-secretase leads to a variety of Aβ species, the major portion (~ 80 - 90%) being of 40 amino acid residues (Aβ40) followed by a 42 amino acid peptide (Aβ42; (5 - 10%), the latter being more hydrophobic and constituting the main deposit in AD brain (Johnson et al., 2002; Dentchev et al., 2003).

Removal of Aβ rom the ECM occurs by transport mechanisms and enzymatic hydrolysis. In Bruch's, Aβ diffuses across the membrane for clearance into the choroidal circulation whereas in the brain, transport is by drainage along perivascular basement membranes and receptor-mediated transport across the blood-brain barrier.

Aβ can be hydrolysed by activated MMPs 2&9 and is the only route for hydrolysis in Bruch's membrane. The brain however houses over 20 multifunctional metallo-endopeptidases that can hydrolyse Aβ and include neprilysin, insulin degrading enzyme, angiotensin converting enzyme, endothelin converting enzyme, and MMPs (Rivera et al., 2010; Chami and Checler, 2012). Despite this armoury, only activated MMP9 was able to degrade fibrillar Aβ in-vitro and compact plaques *in situ* (Yan et al., 2006). The degradation of Aβ by activated MMPs 2&9 is thought to be a central mechanism in amyloid catabolism (Backstrom et al. 1996; Yan et al., 2006; Roher et al., 1994; Yin et al., 2006).

The major Aβ peptides (Aβ40 & Aβ42) are degraded sequentially by MMPs 2&9 to yield three major proteolytic products Aβ (1-34), Aβ (1-30), and Aβ (1-16), the latter being highly soluble is easily removed from the ECM (Hernandez-Guillamon et al., 2015).

Although activated MMP9 could degrade fibrillar Aβ and compact plaques, it showed poor efficacy in the hydrolytic scheme with very little conversion of the intermediate peptides to the final product Aβ (1-16). Activated MMP2 on the other hand rapidly hydrolysed the intact Aβ40 and Aβ42 yielding the final soluble 16 amino acid fragment for clearance. Thus, the combination of activated MMP9/MMP2 would allow the disintegration of amyloid plaques and the hydrolysis of the amyloid species.

As in AMD, despite the presence of pro-MMPs 2&9 in brain, levels of Aβ continue to accumulate in AD providing the initial pathological insult that results in eventual dementia. If the MMP pathway was also present in brain tissue, it could provide a mechanism to better understand the reasons for diminished hydrolysis and clearance of the Aβ deposits.

In a limited study utilising parietal brain tissue from four control and five AD donors, the likely presence of the MMP pathway has been investigated (Hussain et al., 2017). Because of the high lipid content in brain tissue, extraction buffers utilised higher concentrations (4%) of sodium dodecyl sulphate. Tissue samples were homogenised in Tris-buffered saline (TBS, 50mM Tris, 0.15M NaCl, 10mM $CaCl_2$, and 0.02% sodium azide, pH 7.4). They were then mixed 1:1 (v/v) with 4% SDS sample buffer and processed for zymography (Figure 18).

The high molecular weight components, characteristic of the MMP pathway, namely HMW1 and HMW2 were identified in brain tissue. Previous reports have also documented the presence of high molecular components in AD brains and in animal models of AD (Backstrom et al., 1992; Lim et al. 1997).

The likely presence of the LMMC complex in brain tissue was also assessed. For these experiments, the tissue homogenate was spun down and the supernatant applied to gel filtration columns for fractionation. The eluants from two control and two AD donors were then subjected to zymographic analyses (Figure 19). The content of MMP species present in the supernatant of each donor is shown in the left lane labelled Ext. Gel filtration resulted in considerable dilution of the applied supernatant but HMW2 and pro-MMP9 species were observed to be present at the forefront of the chromatographic run (fractions 11-13) indicative of the presence of the LMMC complex.

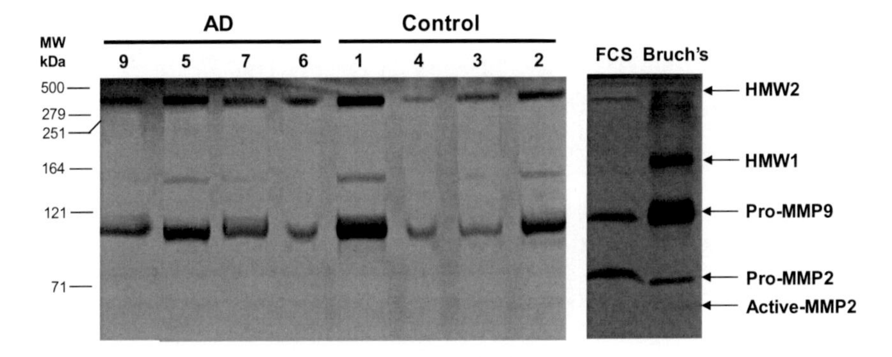

Figure 18. High molecular weight species in brain tissue. Homogenates from four control and 4 AD donors were examined. For comparative purposes, the MMP species present in human Bruch's membrane is also shown.

Zymograms of control donors were dominated by MMP species associated with the LMMC complex but very low or absent levels of free species. On the other hand, zymograms of AD donors showed little presence of LMMC associated MMP species but obvious presence of free MMP species.

The presence of high molecular weight species (HMW1 & HMW2) together with the LMMC complex demonstrate the operation of the MMP Pathway in brain tissue. The status of the MMP pathway in both control and AD donors has also been examined by measuring the level of components in homogenates and supernatants and in the distribution between free and bound compartments (Table 2).

HMW2 levels were similar in both control and AD samples with over 90% of the species being in the bound fraction (similar to Bruch's membrane). The total amount of HMW1 was much lower in AD samples ($p < 0.05$) but levels were similar in the supernatant. The percentage of HMW1 bound was very different compared to Bruch's membrane. In controls, only

47% was bound and this binding was considerably reduced in AD (13%, $p < 0.05$). We do not know if the HMW1 is a complex between pro-MMP9 and NGAL (as suggested in Bruch's membrane); other members of the lipocalin family may be involved leading to the observed changes in the amount bound.

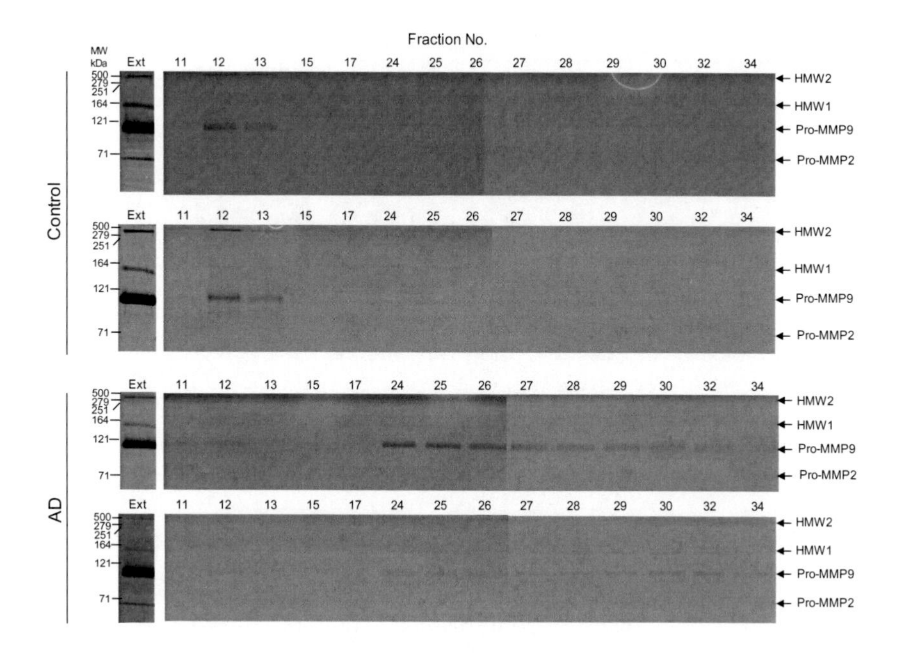

Figure 19. Gel filtration chromatography and subsequent zymographic analyses of brain supernatants from control and AD donors showing the presence of the LMMC complex.

The amount of total pro-MMP9 was similar in both controls and AD samples, but as with HMW1, the amount existing in the bound fraction was reduced (56% and 29% in control and AD samples respectively, $p < 0.05$).

Levels of pro-MMP2 were barely detectable in brain samples and therefore a statistical comparison could not be undertaken. Much work is required to concentrate eluant samples following gel filtration to accurately quantify pro-MMP2 levels.

Table 2. Brain MMP levels and their relative distribution between free and bound compartments

A. MMP activity (gel band area/µg protein) Mean ± SD				
MMP Species	**Homogenate**		**Supernatant**	
	Control	**AD**	**Control**	**AD**
HMW2	35.9 ± 24.2	35.5 ± 7.7	14.7 ± 13.2	10.4 ± 1.8
HMW1	6.4 ± 3.8	1.8 ± 2.0*	8.3 ± 3.2	6.9 ± 5.0
Pro-MMP9	65.7 ± 48.6	51.8 ± 20	108 ± 75	111 ± 37
Pro-MMP2	NDet	NDet	2.33 ± 1.2	2.7 ± 0.4

B. Percentage of bound MMP species (Mean ± SD)		
MMP Species	**Control**	**AD**
HMW2	93.8 ± 5.0	90.3 ± 1.3
HMW1	47.3 ± 24	13.3 ± 14.3*
Pro-MMP9	55.5 ± 11.8	28.5 ± 15.4*
Pro-MMP2	0	0

* $p < 0.05$; NDet: Not detected.

The relative distribution of pro-MMP9 and HMW1 between the LMMC complex and free forms in the supernatants of control and AD donors showed considerable differences (Figure 20). Free levels of pro-MMP9 were low in control supernatants but much elevated in AD donors. The high level of free pro-MMP9 is would favour covalent interaction with pro-MMP2 to generate the HMW2 complex. This is expected to considerable reduce the free level of pro-MMP2 for the activation process and may underlie the reduced degradation of Aβ in AD.

In the supernatant of controls, all the HMW1 species was confined to the LMMC complex but in AD, this confinement was low with much elevated levels being free.

Overall, the MMP pathway has been shown to be present in brain. Unlike Bruch's membrane from AMD donors, the MMP pathway has not been shifted to the left in AD compared to controls. However, the high level of free pro-MMP9 is expected to considerably reduce the free level of pro-MMP2 for the activation process and may underlie the reduced degradation of Aβ in AD. Also, of concern is the high level of free HMW1 in AD tissue and its possible involvement in regulating the MMP pathway requires further investigation.

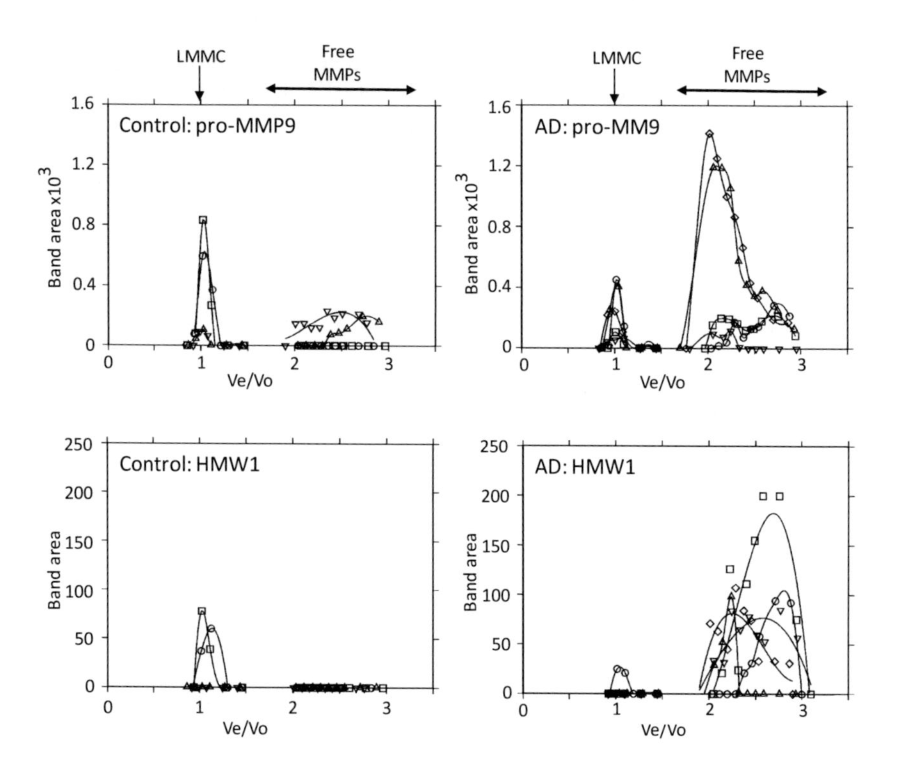

Figure 20. Gel filtration analysis of pro-MMP9 and HMW1 in brain supernatants from control and AD donors. The LMMC complex was present in both control and AD brain samples although the level in AD was reduced. In control brain supernatant, all the HMW1 species was confined to the LMMC complex but in AD, the majority of the species existed in the free form. Similarly, the level of free pro-MMP9 was much higher in AD.

Summary

The complexity inherent in the MMP system arises from the generation of high molecular weight entities and the partitioning between free and bound/sequestered compartments that compromise the availability of the pro-enzymes for the activation process. Further complications arise from the role of TIMPs in regulating the physiological degradation potential of a tissue. In the case of AMD, the availability of free pro-MMP2 for activation was compromised resulting in a 50% reduction in levels of active-MMP2.

Thus, the reduced degradation potential promotes the accumulation of abnormal products in Bruch's leading to the documented demise in structural and functional properties of the membrane.

The MMP pathway was also shown to be present in brain and in Alzheimer's, the presence of large amounts of free pro-MMP9 is predicted to increase the covalent conversion of pro-MMP2 to the HMW2 product, compromising the free level for activation. Since active-MMP2 can effectively degrade Aβ, reduced levels of this enzyme will contribute to the accumulation of this amyloid product in brain. Further work is urgently required to quantify the dynamic aspects of the MMP system in Alzheimer's disease.

REFERENCES

Ahir, A., Guo, L., Hussain, A. A. & Marshall, J. (2002). Expression of metalloproteinases from human retinal pigment epithelium cells and their effects on the hydraulic conductivity of Bruch's membrane. Invest. *Ophthalmol. Vis. Sci.*, *43*, 458-465.

Anderson, D. H., Talaga, K. C., Rivest, A. J., Barron, A. J., Hageman, G. S. & Johnson, L. V. (2004). Characteristics of β-amyloid assemblies in drusen: the deposits associated with aging and age-related macular degeneration. *Exp Eye Res*, *78*, 243-256.

Backstrom, J. R., Miller, C. A. & Tokes, Z. A. (1992). Characterization of neutral proteases from Alzheimer-affected and control brain specimens: Identification of calcium-dependent metalloproteinases from the hippocampus. *J. Neurochem.*, *58*, 983-992.

Backstrom, J. R., Lim, G. P., Cullen, M. J. & Tokes, Z. A. (1996). Matrix metalloproteinase-9 (MMP-9) is synthesised in neurons of the human hippocampus and is capable of degrading the amyloid-β peptide (1-40). *J. Neurosci.*, *16*, 7910-7919.

Bannikov, G. A., Karelina, T. V., Collier, I. E. & Marmer, B. L. (2002). Substrate binding of gelatinase B induces its enzymatic activity in the presence of intact pro-peptide. *J. Biol. Chem.*, *277*, 16022-16027.

Birkedal-Hansen, H., Moore, W. G. L., Bodden, M. K., Windsor, L. J., Birkedal-Hansen, B., DeCarlo, A. & Engler, J. A. (1993). Matrix metalloproteinases: a review. *Crit. Rev. Oral Biol. Med.*, *4*, 197-250.

Butler, G. S., Butler, M. J., Atkinson, S. J., Will, H., Tamura, T., Schade van Westrum, S., Crabbe, T., Clements, J., d'Ortho, M. P. & Murphy, G. (1998). The TIMP-2 membrane type I metalloproteinase 'receptor' regulates the concentration and efficient activation of progelatinase A: a kinetic study. *J Biol. Chem.*, *273*, 871-880.

Chami, L. & Checler, F. (2012). BACE1 is at the crossroad of a toxic vicious cycle involving cellular stress and β-amyloid production in Alzheimer's disease. *Mol. Neurodegener.*, *7*, 52.

Chau, K. Y., Sivaprasad, S., Patel, N., Donaldson, T. A., Luthert, P. J. & Chong, N. V. (2008). Plasma levels of matrix metalloproteinase-2 and –9 (MMP2 and MMP9) in age-related macular degeneration. *Eye* (Lond), *22*, 855-9.

Chin, J. R. & Werb, Z. (1997). Matrix metalloproteinases regulate morphogenesis, migration and remodelling of epithelium, tongue skeletal muscle and cartilage in the mandibular arch. *Development*, *124*, 1519-1530.

Crabb, J. W., Miyagi, M., Gu, X., Shadrach, K., West, K. A., Sakaguchi, H., Kamei, M., Hasan, A., Yan, L., Rayborn, M. E., Salomon, R. G. & Hollyfield, J. G. (2002). Drusen proteome analysis: an approach to the etiology of age-related macular degeneration. *Proc. Natl. Acad. Sci. USA*, *99*, 14682-7.

Curcio, C. A., Presley, J. B., Malek, G., Medeiros, N. E., Avery, D. V. & Kruth, H. S. (2005). Esterified and unesterified cholesterol in drusen and basal deposits of eyes with age-related maculopathy. *Exp. Eye Res.*, *81*, 731-41.

Dentchev, T., Milam, A. H., Lee, V. M., Trojanowski, J. Q. & Dunaief, J. L. (2003). Amyloid-beta is found in drusen from some age-related macular degeneration retinas, but not in drusen from normal retinas. *Mol. Vis.*, *14*, 184-190.

Dufour, A., Zucker, S., Sampson, N. S., Kuscu, C. & Cao, J. (2010). Role of matrix metalloproteinase-9 dimers in cell migration. *J. Biol. Chem.*, *285*, 35944-35956.

Fiotti, N., Pedio, M., Battaglia, P. M., Atamura, N., Uxa, L., Guarnieri, G., Giansante, C. & Ravalico, G. (2007). MMP-9 microsatellite polymorphism and susceptibility to exudative form of age-related macular degeneration. *Genet. Med.*, *4*, 272-7.

Flower, D. R. (1996). The lipocalin protein family: structure and function. *Biochem. J.*, *318*, 1-14.

Guan, K. P., Ye, H. Y., Yan, Z., Wang, Y. & Hou, S. K. (2003). Serum levels of endostatin and matrix metalloproteinase-9 associated with high stage and grade primary transitional cell carcinoma of the bladder. *Urology*, *61*, 719-723.

Guo, L., Hussain, A. A., Limb, G. A. & Marshall, J. (1999). Age-dependent variation in metalloproteinase activity of isolated human Bruch's membrane and choroid. *Invest. Ophthalmol. Vis. Sci.*, *40*, 2676-2682.

Hamlin, C. R. & Kohn, R. R. (1971). Evidence for progressive, age-related structural changes in post-mature human collagen. *Biochim. Biophys. Acta.*, *236*, 458-467.

Handa, J. T., Verzijl, N., Matsunaga, H., Aotaki-Keen, A., Lutty, G. A., te Koppele, J. M., Miyata, T. & Hjelmeland, L. M. (1999). Increase in the advanced glycation end-product pentosidine in Bruch's membrane with age. *Invest. Ophthalmol. Vis. Sci.*, *40*, 775-779.

Hernandez-Guillamon, M., Mawhirt, S., Blais, S., Montaner, J., Neubert, T. A., Rostagno, A. & Ghiso, J. (2015). Sequential amyloid-β degradation by the matrix metalloproteinases MMP-2 and MMP-9. *J. Biol. Chem.*, *290*, 15078-15091.

Holz, F. G., Sheraidah, G. S., Pauleikhoff, D. & Bird, A. C. (1994). Analysis of lipid deposits extracted from human macular and peripheral Bruch's membrane. *Arch. Ophthalmol.*, *112*, 402-406.

Howard, E. W., Bullen, E. C. & Banda, M. J. (1991). Regulation of the autoactivation of human 72-kDa progelatinase by tissue inhibitor of metalloproteinases-2. *J. Biol. Chem.*, *266*, 13064-13069.

Hussain, A. A., Rowe, L. & Marshall, J. (2002). Age-related alterations in the diffusional transport of amino acids across the human Bruch's-choroid complex. *J. Opt. Soc. Am. A.*, *19*, 166-172.

Hussain, A. A., Starita, C. & Marshall, J. (2004). Chapter IV. Transport characteristics of ageing human Bruch's membrane: Implications for AMD. In *Focus on Macular Degeneration Research*, (Editor O. R. Ioseliani). Pages 59-113. Nova Science Publishers, Inc. New York.

Hussain, A. A., Lee, Y. & Marshall, J. (2010a). High molecular-weight gelatinase species of human Bruch's membrane: Compositional analyses and age-related changes. *Invest. Ophthalmol. Vis. Sci.*, *51*, 2363-2371.

Hussain, A. A., Starita, C., Hodgetts, A. & Marshall, J. (2010b). Macromolecular diffusion characteristics of ageing human Bruch's membrane: Implications for age-related macular degeneration (AMD). *Exp. Eye Res.*, *90*, 703-710.

Hussain, A. A., Lee, Y., Zhang, J. J. & Marshall, J. (2011). Disturbed matrix metalloproteinase activity of Bruch's membrane in age-related macular degeneration. Invest. *Ophthalmol. Vis. Sci.*, *52*, 4459-4466.

Hussain, A. A., Lee, Y., Zhang, J. J. & Marshall, J. (2014). Characterization of the gelatinase system of the laminar human optic nerve, and surrounding annulus of Bruch's membrane, choroid, and sclera. *Invest. Ophthalmol. Vis. Sci.*, *55*, 2358-2361.

Hussain, A. A., Lee, Y., Zhang, J. J., Francis, P. T. & Marshall, J. (2017). Disturbed matrix metalloproteinase pathway in both age-related macular degeneration and Alzheimer's disease. *J. Neurodegenerative Diseases.*, 1-13.

Isas, J. M., Luibl, V., Johnson, L. V., Kayed, R., Wetzel, R., Glabe, C. G., Langen, R. & Chen, J. (2010). Soluble and mature amyloid fibrils in drusen deposits. *Invest Ophthalmol. Vis. Sci.*, *51*, 1304-10.

Jackson, G. R., Owsley, C. & McGwin, G. (1999). Aging and dark adaptation. *Vis. Res.*, *39*, 3975-3982.

Jacobson, S. G., Cideciyan, A. V., Regunath, G., Rodriguez, F. J., Vandenburg, K., Sheffield, V. C. & Stone, E. M. (1995). Night blindness in Sorsby's fundus dystrophy reversed by vitamin A. *Nature Genet.*, *11*, 27-32.

Johnson, L. V., Leitner, W. P., Rivest, A. J., Staples, M. K., Radeke, M. J. & Anderson, D. H. (2002). The Alzheimer's A beta-peptide is deposited at sites of complement activation in pathologic deposits associated with aging and age-related macular degeneration. *Proc. Natl. Acad. Sci. USA.*, *99*, 11830-11835.

Kamei, M. & Hollyfield, J. G. (1999). TIMP-3 in Bruch's membrane: Changes during aging and in age-related macular degeneration. *Invest. Ophthalmol. Vis. Sci.*, *40*, 2367-2375.

Karwatowski, W. S. S., Jefferies, T. E., Duance, V. C., Albon, J., Bailey, A. J. & Easty, D. L. (1995). Preparation of Bruch's membrane and analysis of the age-related changes in the structural collagens. *Brit. J. Ophthalmol.*, *79*, 944-952.

Kjeldsen, L., Johnsen, A. H., Sengelov, H. & Borregaard, N. (1993). Isolation and primary structure of NGAL, a novel protein associated with human neutrophil gelatinase. *J. Biol. Chem.*, *268*, 10425-10432.

Kleiner, D. E., Tuuttila, A., Tryggvason, K. & Stetler-Stevenson, W. G. (1993). Stability analysis of latent and active 72-kDa type collagenase: The role of tissue inhibitor of metalloproteinases-2 (TIMP-2). *Biochemistry*, *32*, 1583-1592.

Kumar, A., El-Osta, A., Hussain, A. A. & Marshall, J. (2010). Increased sequestration of matrix metalloproteinases in ageing human Bruch's membrane: Implications for ECM turnover. *Invest. Ophthalmol. Vis. Sci.*, *51*, 2664-2670.

Lee, Y., Hussain, A. A., Seok, J. H., Kim, S. H. & Marshall, J. (2015). Modulating the transport characteristics of Bruch's membrane with steroidal glycosides and its relevance to age-related macular degeneration (AMD). *Invest. Ophthalmol. Vis. Sci.*, *56*, 8403-8418.

Lim, G. P., Russell, M. J., Cullen, M. J. & Tokes, Z. A. (1997). Matrix metalloproteinases in dog brains exhibiting Alzheimer-like characteristics. *J. Neurochem.*, *68*, 1606-1611.

Matrisian, L. M. (1992). The matrix degrading metalloproteinases. *Bioessays.*, *14*, 455-463.

Mott, J. D., Khalifah, R. G., Nagase, H., Shield, C. F., Hudson, J. K. & Hudson, B. G. (1997). Nonenzymatic glycation of type IV collagen and matrix metalloproteinase susceptibility. *Kidney Int.*, *52*, 1302-1312.

Nagai, N., Klimava, A., Lee, W. H., Izumi-Nagai, K. & Handa, J. T. (2009). CTGF is increased in basal deposits and regulates matrix production through the ERK (p42/p44 mapk) MAPK and the p38 MAPK signalling pathways. *Invest. Ophthalmol. Vis. Sci.*, *50*, 1903-1910.

Nagase, H. & Woessner, J. F. (1999). Matrix metalloproteinases. *J. Biol. Chem.*, *274*, 21491-21494.

Nakajima, M., Welch, D. R., Wynn, D. M., Tsuruo, T. & Nicolson, G. L. (1993). Serum and plasma M® 92,000 progelatinase levels with lung cancer metastasis and response to therapy. *Cancer Res.*, *53*, 5802-7.

Owsley, C., McGwin, G., Jackson, G. R., Heinburger, D. C., Piyathilake, C. J., Klein, R., White, M. F. & Kallies, K. (2006). Effect of short term, high-dose retinol on dark adaptation in age and age-related maculopathy. *Invest. Ophthalmol. Vis. Sci.*, *47*, 1310-8.

Park, A. J., Matrisian, L. M., Kells, A. F., Pearson, R., Yuan, Z. Y. & Navre, M. (1991). Mutational analysis of the transin (rat stromelysin) autoinhibitory region demonstrates a role for residues surrounding the "cysteine switch". *J. Biol. Chem.*, *266*, 1584-1590.

Patryas, L., Parry, N. R. A., Carden, D., Baker, D. H., Kelly, J. M. F., Aslam, T. & Murray, I. J. (2013). Assessment of age changes and reproducibility for computer-based rod dark adaptation. *Graefes. Arch. Clin. Exp. Ophthalmol.*, *251*, 1821-1827.

Pilcher, B. K., Wang, M., Qin, X. J., Parks, W. C., Senior, R. M. & Welgus, H. G. (1999). Role of matrix metalloproteinases and their inhibition in cutaneous wound healing and allergic contact hypersensitivity. *Ann. NY. Acad. Sci.*, *878*, 12-24.

Ramratten, R. S., van der Schaft, T. L., Mooy, C. M., de Bruijn, W. C., Mulder, P. G. H. & de Jong, P. T. V. M. (1994). Morphometric analysis of Bruch's membrane, the choriocapillaris and the choroid in ageing. *Invest. Ophthalmol. Vis. Sci.*, *35*, 2857-2864.

Rivera, S., Khrestchatisky, M., Kaczmarek, L., Rosenberg, G. A. & Jaworski, D. M. (2010). Metzincin proteases and their inhibitors, foes or friends in nervous system physiology? *J. Neurosci.*, *30*, 15337-15357.

Roher, A. E., Kasunic, T. C., Woods, A. S., Cotter, R. J., Ball, M. J. & Fridman, R. (1994). Proteolysis of Aβ peptide from Alzheimer disease brain by gelatinase. *A. Biochem. Biophys. Res. Commuun.*, *205*, 1755-1761.

Roomi, M. W., Kalinovsky, T., Rath, M. & Niedzwiecki, A. (2014). Effect of a nutrient mixture on matrix metalloproteinase-9 dimers in various human cancer cell lines. *Int. J. Oncol.*, *44*, 986-992.

Roomi, M. W., Kalinovsky, T., Rath, M. & Niedzwiecki, A. (2015). Failure of matrix metalloproteinase-9 dimer induction by phorbol 12-myristate 13-acetate in normal human cell lines. *Oncology Letters*, *9*, 2871-2873.

Shimarjiri, S., Arima, N., Tanimoto A., Murata, Y., Hamda, T., Wang, K. Y. & Sasaguri, Y. (1999). Shortened microsatellite d(CA)21 sequence down-regulates promoter activity of matrix metalloproteinase 9 gene. *FEBS Lett.*, *455*, 70-74.

Smine, A. & Plantner, J. J. (1997). Membrane type-1 matrix metallo-proteinase in human ocular tissues. *Curr. Eye Res.*, *16*, 925-929.

Springman, E. B., Angleton, E. L., Birkedal-Hansen, H. & Van Wart, H. E. (1990). Multiple modes of activation of latent human fibroblast collagenase: evidence for the role of a Cys73 active-site zinc complex in latency and a "cysteine switch" mechanism for activation. *Proc. Natl. Acad. Sci. USA*, *87*, 364-368.

Starita, C., Hussain, A. A., Pagliarini, S. & Marshall, J. (1996). Hydrodynamics of ageing Bruch's membrane: Implications for macular disease. *Exp. Eye Res.*, *62*, 565-572.

Strongin, A. Y., Collier, I., Bannikov, G., Marmer, B. L., Grant, G. A. & Goldberg, G. I. (1995). Mechanism of cell surface activation of 72kDa type IV collagenase. Isolation of the activated form of the membrane metalloproteinase. *J Biol Chem.*, *270*, 5331-5338.

Triebel, S., Blaser, J., Reinke, H. & Tschesche, H. (1992). A 25 kDa α2-microglobulin-related protein is a component of the 125 kDa form of human gelatinase. *FEBS*, *314*, 386-388.

Upadhya, A. G., Harvey, R. P., Howard, T. K., Lowell, J. A., Shenoy, S. & Strasberg, S. M. (1997). Evidence for a role for matrix metalloproteinases in cold preservation injury of the livers in human and in the rat. *Hepatology*, *26*, 922-928.

Vater, C. A., Harris, E. D. & Siegel, R. E. (1979). Native cross-links in collagen fibrils induce resistance to human synovial collagenase. *Biochem. J.*, *181*, 639-645.

Willenbrock, F., Crabbe, T., Slocombe, P. M., Sutton, C. W., Docherty, A. J., Cockett, M. I., O'Shea, M., Brocklehurst, K., Phillips, I. R. & Murphy, G. (1993). The activity of the tissue inhibitors of the metalloproteinases is regulated by C-terminal domain interactions: A kinetic analysis of the inhibition of gelatinase A. *Biochemistry*, *32*, 4330-4337.

Woessner, J. F. (1995). Quantification of matrix metalloproteinases in tissue samples. *Methods Enzymol.*, *248*, 510-528.

Yan, L., Borregaard, N., Kjeldsen, L. & Moses, M. A. (2001). The high molecular weight urinary matrix metalloproteinase (MMP) activity is a complex of gelatinase B/MMP9 and neutrophil-associated lipocalin (NGAL). Modulation of MMP-9 activity by NGAL. *J. Biol. Chem.*, *276*, 40, 37258-37265.

Yan, P., Hu, X., Song, H., Yin, K., Bateman, R. J., Cirrito, J. R., Xiao, Q., Hsu, F. F., Turk, J. W., Xu, J., Hsu, C. J., Holtzman, D. M. & Lee, J. M. (2006). Matrix metalloproteinase-9 degrades amyloid-β fibrils *in vitro* and compact plaques *in situ*. *J Biol. Chem.*, *281*, 24566-24574.

Ye, S. (2000). Polymorphism in matrix metalloproteinase gene promoters: implication in regulation of gene expression and susceptibility of various diseases. *Matrix Biol.*, *19*, 623-9.

Yin, K. J., Cirrito, J. R., Yan, P., Hu, X., Xiao, Q., Pan, X., Bateman, R., Song, H., Hsu, F. F., Turk, J., Xu, J., Hsu, C. Y., Mills, J. C., Holtzman, D. M. & Lee, J. M. (2006). Matrix metalloproteinases expressed by astrocytes mediate extracellular amyloid-β peptide catabolism. *J. Neurosci.*, *26*, 10939-10948.

In: A Closer Look at Metalloproteinases
Editor: Lena Goodwin
ISBN: 978-1-53616-517-3

Chapter 2

MMP-9 OVEREXPRESSION TOWARDS THE CLINICAL STADIUM OF NASOPHARYNGEAL CARCINOMA

Farhat Farhat[1,*], Elvita Rahmi Daulay[2], Jessy Chrestella[3] and Indah Afriani Nasution[1]

[1]Department of Otorhinolaryngology Head and Neck Surgery, Faculty of Medicine, Universitas Sumatera Utara, Medan, Indonesia
[2]Department of Radiology, Faculty of Medicine, Universitas Sumatera Utara, Medan, Indonesia
[3]Department of Anatomic Pathology, Faculty of Medicine, Universitas Sumatera Utara, Medan, Indonesia

ABSTRACT

Nasopharyngeal carcinoma (NPC) is one of the most predominant cancers in Southern China and Southeast Asia. Matrix metalloproteinase (MMP)-9 has played several crucial roles in carcinogeneses, such as tumor invasion, metastases, and tumor vascularization. Extracellular matrix degradation has been linked to aggressive tumor invasion into the surrounding tissue as well as vascular or lymphatic vessels. This study aims

* Corresponding Author's E-mail: farhat@usu.ac.id.

to evaluate the expression of MMP-9 in nasopharyngeal carcinoma patients. The cross-sectional analytic study enrolled 106 patients with nasopharyngeal carcinoma based on radiologic and histopathological examination. The patient never underwent radiotherapy, chemotherapy and/or a combination. MMP-9 was assessed using immunohistochemical staining of MMP-9 in the Anatomic Pathology Department. In this study, most NPC patients were found in the age group of 41-60 as much as 72 patients (56.6%). Most NPC patients were males as much as 90 patients (69.8%). The histopathology showed the most overexpression was found in non-keratinizing squamous cell carcinoma (80.5%), Primary Tumor T3 (35.4%), Nodes N3(56.1%), and stage IV (70.7%). Through an analysis study using Fisher's exact test, it was found p value < 0.001, which showed a significant correlation between MMP-9 overexpression to the clinical stage of NPC. It suggests the evidence that MMP-9 must be unveiled its potency as a biomarker of nasopharyngeal caricnoma. The high expression of MMP-9 in nasopharyngeal carcinoma patients supported the evidence of its role in the disease and so the potentiation of the inhibitor of MMP-9 as targeted therapy of nasopharyngeal carcinoma could be beneficial for the disease.

Keywords: matrix metalloproteinase, clinical stage, invasion, histopathology

INTRODUCTION

As one of the most common malignancies found in head and neck areas, nasopharyngeal carcinoma has distributed differently worldwide. There is a tendency that nasopharyngeal carcinoma occurs predominantly among certain geographic areas and ethnic groups. Therefore, global cancer registry has ranked nasopharyngeal carcinoma as the 11th most common cancer in the region with an incidence of 20-40 per 100,000 people. In accordance with the continent, 81% of new cases were found in Asia while only 6% in Africa, the rest blatantly spread out across the nations. Specifically, Southeast Asia countries have served as the place in which 67% of NPC patients are diagnosed. In General, the unbalanced prevalence occurs among the population that concentrate in Southeast Asia. Five countries with the highest incidence of NPC are China, Indonesia, Vietnam, India, and Malaysia. In Indonesia, NPC is the 4th most common malignancy after cervical, breast and skin cancer. The average prevalence of nasopharyngeal carcinoma in Indonesia is 6.2 per 100,000 people or about 12,000 new cases of nasopharyngeal carcinoma per year.

Based on recent studies, it has been proved that there is no single causative agent could induce tumorigenesis in NPC patients. In other words, the precise causative agent is still elusive. Whereas, provisional conjectures that may cause NPC such as Ebstein Barr virus infection (EBV), genetic factors and environmental conditions is emphasized in literature. EBV as the most involved predisposing factor in initiating NPC tumor development has been characterized by the presence of mini circular chromosome, called episomes, and the vast majority of the nucleus of malignant cells conceived approximately 30 types of the EBV gene. Therefore, a researcher recently acknowledges that high levels of IgA antibodies against EBV could determine the high likelihood of a person suffers from NPC, in other words, it is a reliable screening tool particularly in high incidence region. Salted preserved fish and several types of foods are also linked to the increase of the relative risk of NPC among the Chinese population, 1.4 to 3.2 and 1.8 to 7.5 among the consumer group. Furthermore, for stratification (clinical stage) and prognostic purposes, American Joint Committee on Cancer (AJCC) 7th edition is applied, it is also used to determine the proper management among patients.

NPC is not a typical squamous malignancy compared to other types of malignancy in head and neck areas. Its location has brought several clinical significance and always constraint the early findings. Therefore, making diagnosis becomes a challenge, since NPC does not show any symptoms and it causes the ignorance of the suferrers. Subsequently, most of NPC patients came to the hospital with advanced stages, metastasis, local or distant, and node involvement. NPC diagnosed usually found in stage III or IV. Finally, it needs more aggressive treatment and the satisfactory results seem unattainable. Until recent days, medical practitioner still uses clinical examination and imaging studies to diagnose NPC. Inspection using fiberscope is the mainstay of the diagnostic approach for nasopharyngeal abnormalities. In addition to the former method, the serologic marker has been extensively used to make the early detection of NPC possible. IgA antibody titers, such as EBV-IgA-VCA (against viral capsid), and EBV-EA (Early antigen of EBV) has placed as the serologic screening tools. In fact, by monitoring EBV-IgA antibody levels prospectively could determine tumor remission and relapse. Previously, literature proves that the surge of IgA antibody precedes the tumor findings. In other studies, when EBV gene was found in the nasopharyngeal swab, indeed, it could predict symptomatic NPC.

In targeted cancer therapy era, the immunohistochemical based examination has been extensively studied. Several basic substances expression has been linked to the more aggressive type of tumor and also the treatment determination. NPC studies and management has followed the trends. Several basic enzymes, substances, and molecules, such as matrix-metalloproteinase, cyclooxygenase, and protein glycosylation, have been associated with NPC diagnostic, treatment modalities and prognostic improvement. Studies have proposed several basic mechanisms related to MMP inducing NPC tumor development. Matrix-metalloproteinases (MMPs) are substance included in a family of zinc-dependent proteinases proteases that function to degrade extracellular matrix component and were found in the living cell. For instance, MMP-9 expresses, specifically, in various type of cell, such as epithelial cells, fibroblasts, endothelial cells as well as inflammatory cells. It primarily acts as tumor and metastasis promoting substances by inducing tumor invasion and angiogenesis, as its capability degrading. On the basis of physiologic purposes, MMPs has certain part in tissue regeneration and wound repair. Later, the finding of MMPs role in carcinogenesis shift researcher's view into a novelty in which MMPs has projected to induce massive cancer cell growth, wrong direction differentiation, ineffective apoptosis, distant migration, and invasion. Later on, it produces tumor metastasis.

MMPs are now targeted for its role on every part during tumor progression and development. It was found that there are 26 MMPs, whether or not it shares common structural and functional similarities. It is well-documented that MMP-2, MMP-3, and MMP-9 as the three most important members of the MMPs family. MMP-2 or gelatinase A could digest and degrade type IV collagen as well as growth factor-binding protein and growth factor receptors, and it is encoded by chromosome 16q13-q21. MMP-3 is useful for disrupting basal membrane collagen, but it also activates MMP-1 and MMP-9 expression. While MMP-9 (Gelatinase B) encoded by chromosome 20q13.12, besides its complex domain structure, can degrade many materials including decorin, elastin, fibrillin, laminin, gelatin, and types IV, V, XI, and XVI collagen. Therefore, overexpression of MMP-2, MMP-3, and MMP-9 is highly related to more advanced and aggressive tumor development. Other than that, they also could found in polymorphism state thus increasing the susceptibility of the individual suffers from NPC. Nevertheless, studies act incessantly to reveal the fixed relationship between the overexpression and tumor which represent by the clinical stage. In addition, MMP-9 complete structure has also not been known yet and still become enigmatic for decades, and MMP function remains inconsistent in many studies.

Human MMP-9 major sources are from neutrophils, macrophage, fibroblasts, and endothelial cells. Firstly, it is synthesized as pre-proenzyme including 19 amino acid N-terminal signal peptides in its cell. In order to be active, MMP-9 must be cleavaged by some proteases, such as MMP-3, by which mechanism aims to remove the N-terminal propeptide region. In addition, proteolytic removal is also affected by reactive oxygen species (ROS) and nitric oxide. ROS will severely impair the activation cascade of MMP-9, thus inducing uncontrollable activation. MMP-9 has been widely investigated for its role in several biological processes, particularly in carcinogenesis. Major negative aspects of MMP-9 in supporting cancer development is related to its role in ECM remodeling and membrane protein degradation. Wide range of substrate makes MMP-9 more vulnerable in affecting tumor microenvironment hence creating a more aggressive and violent tumor. Later it is mentioned that MMP-9 also associated with cancer pathologies.

Tumor development and its progression have been linked with the massive expression of MMP-9 among certain malignancies. Type IV collagen primarily degrades by MMP-9, and the collagen is abundant in the extracellular matrix (ECM). This act could indicate MMP-9 as metastatic inducer. In inflammatory conditions, some malignant cell can induce interleukin-6 in its microenvironment. Subsequently, it can induce MMP-9 overexpression, but the accurate mechanism of how IL-6 could induce MMP-9 expression has not been elucidated yet. As mentioned earlier, immune cell infiltrates take some parts in tumor development by secreting inflammatory cytokine, such as TNF-α and IL-6, reactive oxygen species (ROS) and nitric oxide (NO) indirectly increase the expression of MMP-9. It emphasis that MMP-9 roles have particularly benefitted the tumor cell to carry out metastasis. NF-κB transcriptional activity states as its inducer, at the same time EBV and LMP1 oncogene triggers NF-κB activation. The other vicious cycle provide by certain agents, including NOS and ROS, they can disrupt a cysteine-zinc bond and finally, it will degrade the MMP-9 domain. Therefore, NF-κB, IL-6, NO, and ROS has formed a multistep process that may directly affect MMP-9 activities. In recent studies, among undifferentiated NPC, low survival rates are in relation to the high accumulation of pro-MMP-2 and MMP-9. MMP-9 also impair angiogenesis process as well as initiating metastasis. In the previous study, transgenic mouse model (RIP1-Tag2) was used to define multistage carcinogenesis, after developing islet tumors of the pancreas. Angiogenesis induction occurs as a discrete pathway producing hyperproliferative islet. It was characterized by endothelial proliferation, mitosis, micro-hemorrhaging, and vascular dilation. Conventional hypotheses claimed the angiogenesis state is mostly caused by VEGF, but the study proved other findings that MMP has intermingled in this role with VEGF. The investigation starts examining angiogenic response after administration of MMP-9 and at the same time, islet culture was embedded with anti-VEGF antibodies. The angiogenic response was blocked. It is concluded that MMP-9 mobilized VEGF from normal islets. Therefore, MMP-9 was stated as VEGF inducer, an important regulator of angiogenesis, and finally, it can cause vascular pericyte recruitment. In addition, the study also discovered that mice that were homozygous-null for MMP-9 or MMP-2 are found to have reduced tumor burdens. Furthermore, many ECM and plasma surface proteins could be present, and it can cause activation of certain transcriptional factors that were overwhelmingly inducing other protein expressions until it produces distant metastasis. Distant metastasis is supported by the fact that MMP-9 selectively works degrading the type IV collagen which is a major component of extracellular matrix.

MMP regulate protein at the cell surface, but it might be more complicated since the membrane cell is composed of lipid that consists of protein in bilayer coating. There is 'rafts' found on the outside of the cell membrane, and it controls membrane protein-protein interaction. Rafts consist of several basic components such as glycosphingolipid (GSL), sphingomyelin and cholesterol in a package in the plasma membrane. Rafts role are associated with its activation and inhibitors of MMP. One study proves MMP membrane in tumor progression is localized by generating rafts and non-rafts MMP-9 using the cell line cultures of breast carcinoma. It is concluded from the study that cultures which secreted wild-type MMP-9 caused angiogenic switch while non-raft samples release MMP-9 with physiologic activity. It means that stromal secretion of MMP-9 is the utmost causative factor in increasing capillary number and vessel perimeter. Wild-type MMP-9 were then associated with high expression of VEGFR2, VEGF receptor. In fact, the receptor is also dependent on MMP-9 function. In other studies, neovascularization roles of MMP-9 are evident among herpes simplex virus (HSV) infection of the cornea patients. As neutrophil secreted substance, MMP-9 degrades extracellular matrix facilitating neovessel growth.

Proteolysis of ECM causes cell motility which promotes distant tumor metastasis and invasion. N-cadherin has suggested playing a part for the epithelial-to-mesenchymal transition (EMT). EMT is needed to produce more invasive tumor. N-cadherin is a cell adhesion molecule located in transmembrane, but its high expression is a sign of EMT which related to malignancy and distant metastasis. N-cadherin roles in tumor progression emerge since it profoundly can generate tumor cell survival, migration, and invasion. As the post-translating process, it has been reported that a patient with a high expression of N-cadherin more likely produces poor prognosis. MMP cleavage cadherin resulting in the exposure of the extracellular N-terminal amino fragment (NTF) and first C-terminal fragment (CTF1) which appears in the cytoplasmic side. In further process CTF1 will produce CTF2. The main function of CTFs is primarily as regulatory protein related to cell migration and invasion. In other words, tumor progression in NPC and MMP-9 is related via increasing the N-cadherin level. It is proved that its cleavage is correlated with MMP-9 overexpression and vascular muscle cell proliferation. It was suggested that MMP-9 proteolytic activity caused shedding of extracellular and intracellular fragments. Several significant signaling pathways related to N-cadherin initiate β-catenin and p120-catenin pathways, as well as protein kinase C (PKC)-mediated ADAM10 which leads to glioblastoma cell migration. N-cadherin cleavage also has positive feedback in causing the increase of MMP-9 expression. Thus, unbalanced secretion between these two components is an essential part of inducing aggressive tumor cells.

In oral carcinoma, elevated MMP-9 expression was found in tissue, serum, and saliva samples among oral potentially malignant disorders (OPMDs). Furthermore, the diagnostic value of MMP-9 overexpression is evident. In case of apoptosis, it is then defined that after MMP-9 inactivates the Natural Killer cells, it cleaves the FAS receptor producing tumor with apoptosis-resistant characteristics and it secretes differently in accordance with tissue environment, so-called as an inducible enzyme. While, ovarian cancer and MMP-9 also linked through unique transcription factor, called STAT, and it enrich MMP-9 gene promoter. The poorest prognosis type of breast carcinoma, triple negative, also adhere to the basis of MMP-9 expression drives invasiveness, metastasis, and vessel formation.

MMP-9 could be an important target in eradicating malignant cells and other MMP-9 related diseases. By finding MMP-9 inhibitor could trigger some specific therapies, or monoclonal antibody, as it also increases the effectivity while reducing adverse effects. In future perspective, MMP-9 could be used as cancer biomarkers to predict remission/relapse, monitoring, treatment effectiveness, or malignancy screening, yet, There are still no studies which provide single biomarker could produce high accuracy. Therefore, using combination biomarker can be considered . Our study aimed to analyze and assess MMP-9 overexpression among newly diagnosed NPC patients in one single institution.

METHODS

Sample Collection

The samples were 106 paraffin block of nasopharyngeal carcinoma patients who had not received the chemotherapy and/or radiotherapy obtained at Otorhinolaryngology Head and Neck Surgery Departement, Adam Malik General Hospital Medan, Indonesia. CT scan with contrast was done and the diagnosis was made based on histopathological biopsy. Then, the paraffin block was cut into section and being processed with immunohistochemistry staining.

Samples was collected using the consecutive sampling method. The patients with other malignancies or damaged paraffin block were excluded from this research. This research was conducted from July to October 2017 and the immunohistochemistry staining was done in Anatomy Pathologic Department, Faculty of Medicine, Universitas Sumatera Utara.

The data were grouped in accordance with gender, age, histopathological examination results and clinical stage (AJCC, 2010). AJCC criteria for clinical stage consisted of primary tumor size (T), lymph node involvement (N), and metastasis (M). Ethical clearance approval had been obtained from the Health Research Ethics Committee of The Faculty of Medicine, Universitas Sumatera Utara. The committee also declared that the study was in line with Helsinki declaration for human studies.

Immunohistochemistry

Deparaffinization of NPC samples was firstly performed before its processing into immunohistochemical staining in 100% xylene. Then, serial alcohol rehydrates the samples using 96%, 80%, and 70% absolute alcohol and it was performed according to the given instruction, followed by heating antigen causing retrieval, set up preheat in 65^0C, and running time in 98^0C for 15 minutes. To promote inhibition of endogenous peroxidase and nonspecific substance, the specimen was administered with a peroxidase blocking agent, 3% hydrogen peroxide, while 3% Normal Horse Serum (NHS) used to block other unknown antigens. After that, the sepcimen was rewashed using Triss Buffered Salin (TBH), and it was incubated with goat antihuman MMP-9 antibody (1:100) at 37^0C for one hour. Second rewash similarly conducted, the specimen was embedded once again using Dako Real Envision Rabbit/Mouse for 30 minutes. Peroxidase resulting in the addition of 3.3-diaminobenzidine chromogen solution mixed with DAB buffer solution (20 μL DAB: 1000μL substrate). Lastly, the application of counterstaining, using hematoxylin, was done before it was examined under the microscope light.

Staining Evaluation

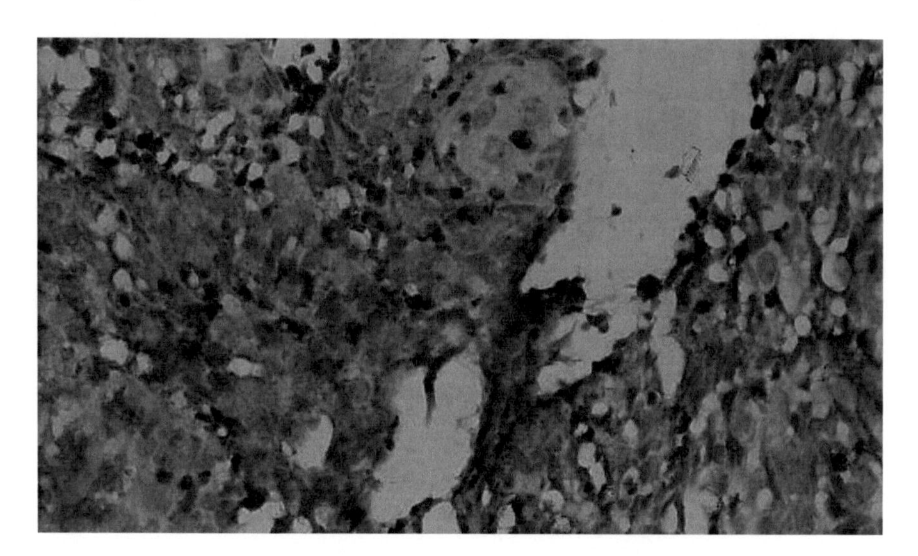

Figure 1. Cytoplasmic expression of MMP-9 in non-keratinizing squamous cell nasopharyngeal carcinoma (x 400).

The evaluation was carried out by two pathologists who did not know about the clinical and tumor characteristics of the patients. The pathologist assessed the immunohistochemical staining by determining each immunoreactive scores. The scores were obtained after two different value of intensity (0-3 from negative until strong) and a wide area of staining (0 negatives, 1 for positive staining beyond 10% of cell count, 2 for 10-50% of cell count, 3 for positive in more than 50% cell count) being multiplied. The final result showed greater than or equal to 4 termed as 'overexpression'. MMP-9 expression was assessed by using histopathological tests based on immunoreactivity (negative = 0-3, positive/overexpression = 4-9). All data are presented in the table.

Statistical Analysis

The data were processed using Statistical Package for Social Sciences (SPSS) 23.0 software. Univariately, the demographic data and clinical characteristics of the patients were depicted in the single table in order to reveal p-value of the MMP-9 overexpression and clinical stadium correlation, the analysis was performed using Fisher's exact test. If a p-value of less than 0.05, the results were considered statistically significant.

RESULT

This study involved 106 samples of paraffin block nasopharyngeal carcinoma patients with the most age group is 41-60 years as much as 60 samples (56.6%), while in the gender category found that the male group is the largest population of NPC patients as much as 74 samples (69.8%). In the histopathologic group, the most type was non-keratinizing squamous cell carcinoma as much as 84 samples (79.2%). While the most common for primary tumor was T3 (32.1%) and lymph node metastasis was N3 (44.3%). The clinical stadium IV was the most stadium found in nasophayngeal carcinoma patients (55.7%). In immunohistochemical evaluation, most of MMP-9 overexpressed in non-keratinizing squamous cell carcinoma (80.5%), in primary tumor T3 (35.4%), neck enlargement in N3 (56.1%). Moreover, at the clinical stage, stage IV group (70.7%) was the highest category.

According to the study analysis with Fisher's exact test was obtained p-value <0.001, which indicate that there are significant correlations of MMP-9 overexpression to the clinical stadium of NPC.

Table 1. Demography and Clinical Characteristic of Nasopharyngeal Carcinoma Patients

Age (years)	Characteristic	N	%
	≤ 20	5	4.7
	21 – 40	23	21.7
	41 – 60	60	56.6
	> 60	18	17.0
Sex			
	Male	74	69.8
	Female	32	30.2
Histopathological Type			
	Keratinizing Squamous Cell Carcinoma	8	7.5
	Non-Keratinizing Squamous Cell Carcinoma	84	79.2
	Undifferentiated Carcinoma	14	13.2
Primary Tumor			
	T1	20	18.9
	T2	27	25.5
	T3	34	32.1
	T4	25	23.6
Lymph Node Metastasis	N0	21	19.8
	N1	17	16.0
	N2	21	19.8
	N3	47	44.3
Clinical Stadium			
	Stage I	13	12.3
	Stage II	12	11.3
	Stage III	22	20.8
	Stage IV	59	55.7

Table 2. MMP-9 overexpression in accordance with its clinical characteristics

Characteristic	MMP-9 Expression			
	Overexpression	%	Negative	%
Histopathological type				
Keratinizing squamous cell carcinoma	7	8.5	1	4.2
Non-keratinizing squamous cell carcinoma	66	80.5	18	75.0
Undifferentiated carcinoma	9	11.0	5	20.8
Primary tumor size				
T1	9	11.0	11	45.8
T2	19	23.2	8	33.3
T3	29	35.4	5	20.8
T4	25	30.5	0	0.0
Lymph node metastasis				
N0	9	11.0	12	50.0
N1	10	12.2	7	29.2
N2	17	20.7	4	16.7
N3	46	56.1	1	4.2
Clinical Stadium				
I	4	4.9	9	37.5
II	4	4.9	8	33.3
III	16	19.5	6	25.0
IV	58	70.7	1	4.2

DISCUSSION

The highest incidence of non-keratinizing squamous cell carcinoma was evident from the study, and it was also correlated with MMP-9 overexpression. Overexpression was seen among the rest of two histopathology types of nasopharyngeal carcinoma, undifferentiated carcinoma (11.0%) and keratinizing SCC (8.5%) respectively. Based on American Joint on Cancer Committee 2010 definition for TNM staging, it was found that primary tumor size T3-T4, lymph node involvement N3, and clinical stage IV had more MMP-9 overexpression compared to an early type of tumor. It can be concluded that late stage tumor has an association between several clinical parameters and MMP-9 expression. The study became supportive of the fact that MMP-9 has roles in tumor progression since it was commonly found among advanced-stage tumor. Liu et al. found that the expression level of MMP-9 mRNA increased in NPC samples compared to normal nasopharyngeal tissue. The study also obtained the conclusion that MMP-9 expression may become an independent prognostic indicator for NPC patient survival rate. Based on the previous studies, it is found that the levels of MMP-9 evidently increase among the patient with nasal NK/T-cell lymphoma, malignant astrocytoma, carcinomatous meningitis, and metastasis in various types of the malignant tumor. Expression of MMP-9 accelerate all the processes related to advancing tumor growth and progression. Therefore, the study aimed to provide the information about its presence in tumor progression by relating it with various clinical parameters such as primary tumor size, lymph node involvement, and metastasis, as the effective way to determine disease progression among malignancy. In addition, among the general population, NPC tends to occur among elderly patients. Puspitasari (2011) and Pua et al. (2008) found that the highest incidence of nasopharyngeal carcinoma was among 50-60 years old. The incidence of NPC also increase in the age > 30 years and the peaks of age between 50-59 years and decrease after age > 60. While, this study also proved that major NPC patients was taken place among 41-60 years old patient, as much as 72 samples (57.1%), and had male predominance (71.4%). Xiao et al. found the nasopharyngeal carcinoma patients with the comparison of male and female with a scale of 2.5:1.

The majority of non-keratinized SCC of NPC had MMP-9 overexpression (59.5%). MMPs, particularly MMP-9, are upregulated among head and neck squamous cell carcinoma (HNSCC), not only NPC. Nevertheless, the tendency to overexpress in one histopathology type has not been elucidated. High level of MMP-9 occurs secondarily along with the increase of latent membrane protein 1 (LMP1). Therefore, LMP1 drives certain MMP activity as well as down-regulated CD99 which is well known as the anti-NPC immune response. High-MMP activity in NPC patient is responsible for LMP1 as it becomes the driving force of MMP activity particularly MMP-1 whereas MMP-9 is characteristically more EBV dependent. Another substance which could increase and has an association with MMP-9 expression is protease-activated receptor 2 (PAR-2). In cooperation with LMP1, it was found that poorer prognosis obtained from the patients with dual substance expression both PAR-2/LMP1 and PAR-2/MMP-9. Furthermore, intratumoral microvessel density also showed related to the expression of PAR-2 and LMP1 expression, therefore, it stated that PAR-2 could promote lymph node metastasis.

Recent studies was found double-edge sword function of MMPs among oral cancer, it has some protective mechanism to carcinogenesis, but several studies still found its expression is highly related to the advanced-stage tumor. MMP function is focused on the interaction between cancer cells and stromal cells. It proposed communication between the two component conducted by several growth factors, hormones, cytokines, and proteinases including MMPs resulting in metastasis. One study proves that there was a relationship between carcinoma-associated fibroblasts (CAFs) in oral squamous cell carcinoma cell line inducing expression of MMP-9 via a fibronectin-integrin αvβ6 pathway since its active form was found in aggressive human tongue squamous cell carcinoma cell line HSC-3. In addition, other bioactive substrates that may modulate carcinogenesis are also affected by MMP-9 including pro-transforming growth factor-β1 (TGF-β1) and the pro-tumor necrosis factor-α (TNF-α). Adenocarcinoma-type carcinoma also linked to MMP-9 overexpression, such as among lung, prostate, and breast carcinoma. One study also uncovered there is imbalance between the MMP-9 (overexpression) and TIMP-1 (down-expression) among prostate adenocarcinoma patients. Epithelial cancer, as well as nasopharyngeal carcinoma, is formed in a special microenvironment. Chronic inflammation could change the environment by transforming innate immune cells activities, such as mast cells, granulocytes, and macrophages. They secrete different cytokines, interleukins, reactive oxygen species, and MMPs that could modulate angiogenesis, tumor proliferation, growth, and its invasion. Finally, transforming growth factor β1 become overexpressed, and it enhances snail expression as a transcription factor activating MMP-9 expression and triggers an EMT. In the end, malignant cells will change its morphology and existed in damaged intercellular and cell-matrix adhesion thus inducing cell motility. Activated MMP-9 can also strengthens the vicious cycle including TNF-α, CXCL1, CXCL4, CXCL7, CXCL8 and interleukin-1β which can induce MMP-9 activation. One theory has been proposed that interleukin-1β which secreted by tumor cells will induce lipocalin 2 secretion. Association among lipocalin 2, MMP-9 and its complex is evidently affecting clinical stages and tumor size among oral malignant lesion patients. CXCL8 is also suggested to increase MMP-9 secretion from tertiary neutrophil granules, and its levels may be related to the disease progression. Unfortunately, the mRNA expression of CXCL8 is also related to the activities of TNF-α and IL-1β. Main impacts of CXCL8 in tumor microenvironment include cell migration induction, invasion, and MMP-7 expression, while there is no evidence that CXCL8 could affect MMP-9 expression.

Invasion and tumor progression involves a multistep process. Various type of protein also proposed to acknowledge MMP-9 as an inducer of the process. Casp12 was stated as a regulator of MMP-9 resulting in NPC cell invasion, while CCL2/CCR2 axis provokes cell mobility by increasing MMP-9 expression. Ruokolainen et al. (2004) found shortened relapse-free, and it was because of MMP-9 overexpression found in the majority of the sample, and it affected tumor progression. As the sign of metastasis, lymph node involvement also was highly found among N3 in this study. Wu et al. (2013) also discovered similar results that lymph node metastasis was associated with lymphangiogenesis as well as MMP-9 overexpression among breast cancer patients.

Furthermore, Pryczynicz et al. (2007) proved that significant correlation between MMP-9 expression and node involvement and distant metastasis, and also poor prognosis had been obtained from the study among MMP-9 overexpression and pancreatic ductal carcinoma patients. One cross-sectional study evaluates the expression of receptor for advanced glycation end products (RAGE) and reversion inducing cysteine-rich protein with Kazal motifs (RECK), it found that dysregulation of RAGE and RECK protein can predispose to high MMP-9 expression besides its positive correlation with poor clinicopathological properties in addition to EBVCA-IgA, lymphatic metastasis, recurrence, and survival. However, RECK expression is inversely correlated with RAGE and MMP-9 levels in NPC tissues.

EBV as an emerging causative agent also correlated with MMP-9 expression. MMP-9 overexpression firstly introduced as a result of NF-κB and activator protein (AP)-1 activation by LMP1, later it was found that many antigens and antibodies secreted or encoded by EBV related to carcinogenesis via MMP-9 overexpression. Tang et al. (2004) showed that the expression of EBER-1 had a significant relationship with MMP-9 expression, at the similar occasion, the study also stated that lymph node metastasis is associated with high expression of MMP-9. The conclusion finally demonstrated that EBV can augment metastasis via inducing MMP-9 up-regulation since MMP-9 expression rate found in 73.17% of NPC specimens. Murono et al. (2000) demonstrated that by using Aspirin could inhibit MMP-9 particularly if the tumors express LMP1, consequently it reduced tumor growth and invasion in EBV-associated tumor samples. LMP1 is originally known as an oncoprotein. It is known that LMP1 induces epidermal growth factor receptor (EGFR) expression in the early event of carcinogenesis. It also indirectly upregulates p16 and MMP-9 through upregulating Vascular Endothelial Growth Factor (VEGF) and TGF-α.

In this study, the most common MMP-9 expression encountered was in N3, 55 samples (43.7%). Consistently, Liu et al. (2010) found that there were 78 samples (86.7%) with overexpressed MMP-9 in N3 and Jian et al. (2004) reported there were 32 samples (78.6%) out of 41 samples that occurred as N1-N3. In this study, overexpression of MMP-9 encountered as much as 67 samples (53.2%) in stage IV. This is in accordance with the research result of Liu et al. (2010) that found most of the MMP-9 overexpression in patients with nasopharyngeal carcinomas in stage III-IV. And similar to Puspitasari (2011) that found MMP-9 overexpression in nasopharyngeal stages III-IV carcinoma of 9 samples. (90.0%). However, all fourth clinical stages had been abundant with MMP-9 expression (I-II: 86.4% and III-IV: 73.7%) in a study conducted by Zhou et al. (2014). It brings the evidence that MMP-9 has been implicated in some parts of the tumorigenesis of NPC, particularly among advanced stage tumor. It increases the suspicion that several invasive carcinomas might be indirectly or directly linked to the MMP-9 overexpression, such as colorectal cancer, gastric carcinoma, pancreatic carcinoma, oral cancer, and breast cancer. MMP-9 involvement in promoting the formation of invasive squamous cell carcinoma of the uterine cervix is evident, its expression markedly, and positive correlation with invasion and metastasis was proved among 65 cases compared with the normal cervical epithelium of chronic cervicitis.

Conversely, Akdeniz et al. (2013) discovered that E-cadherin could predict the differentiation and nodal metastases among laryngeal squamous cell carcinoma patients while MMP-9 as a negative predictor of differentiation. The study found that intercellular junction protein has a bigger part causing tumor progression. In melanoma malignant patients, ten patients with metastasis demonstrated larger area with MMP-9 staining and the majority of epithelioid type tumor expressed MMP-9 in contrast to spindle type cell but only MMP-2 and MMP-14 expression have a significant relationship with the patient's outcome.

CONCLUSION

According to the results, the study found an elevated of MMP-9 expression predominantly in non-keratinizing squamous cell carcinoma type, primary tumor T3, nodes N3, and clinical staging IV. In addition, using Fisher's exact test, we found that $p < 0.001$ that suggested there is a relationship between overexpression of MMP-9 to the clinical staging of NPC. Future directions of targeted therapy have been linked to more advantageous approach, particularly among NPC patient which commonly diagnosed in advanced or late stage. Nevertheless, predicting NPC histopathology type in accordance with MMP-9 overexpression is essential part to tailor management which is also related to its histopathological type. Therefore, an appeal to perform prospective and cohort studies are mandatory to show effectiveness between immunohistochemical targeted MMP-9, and is beneficial rather than costly diagnostic approach. The study provides a significant amount of data which can be stated as one of the studies in Indonesia which proves the relationship of MMP-9 overexpression with clinical stage. In addition, some histopathological indicators were not included in the data and predicted to have more effects.

REFERENCES

Adham, Marlinda, Antonius N Kurniawan, Arina Ika Muhtadi, Averdi Roezin, Bambang Hermani, Soehartati Gondhowiardjo, I Bing Tan, and Jaap M Middeldorp. "Nasopharyngeal Carcinoma in Indonesia: Epidemiology, Incidence, Signs, and Symptoms at Presentation." *Chinese journal of cancer* 31, no. 4 (2012): 185.

Akdeniz, Onder, Davut Akduman, Mehmet Haksever, Haluk Ozkarakas, and Bahar Muezzinoglu. "Relationships between Clinical Behavior of Laryngeal Squamous Cell Carcinomas and Expression of Vegf, Mmp-9 and E-Cadherin." *Asian Pac J Cancer Prev* 14, no. 9 (2013): 5301-10.

Andisheh-Tadbir, Azadeh, Maryam Mardani, Sara Pourshahidi, Kamran Nezarati, and Parisa Bahadori. "Prognostic Value of Matrix Metalloproteinase-9 Expression in Oral Squamous Cell Carcinoma and Its Association with Angiogenesis." *Journal of clinical and experimental dentistry* 8, no. 2 (2016): e130.

Babichenko, Igor I, Mikhail I Andriukhin, Sergey Pulbere, and Artem Loktev. "Immunohistochemical Expression of Matrix Metalloproteinase-9 and Inhibitor of Matrix Metalloproteinase-1 in Prostate Adenocarcinoma." *International journal of clinical and experimental pathology* 7, no. 12 (2014): 9090.

Bergers, Gabriele, Rolf Brekken, Gerald McMahon, Thiennu H Vu, Takeshi Itoh, Kazuhiko Tamaki, Kazuhiko Tanzawa, et al. "Matrix Metalloproteinase-9 Triggers the Angiogenic Switch During Carcinogenesis." *Nature cell biology* 2, no. 10 (2000): 737.

Cao, Su-Mei, Malcolm J Simons, and Chao-Nan Qian. "The Prevalence and Prevention of Nasopharyngeal Carcinoma in China." *Chinese journal of cancer* 30, no. 2 (2011): 114.

Chan, KC Allen, Emily CW Hung, John KS Woo, Paul KS Chan, Sing-Fai Leung, Franco PT Lai, Anita SM Cheng, et al. "Early Detection of Nasopharyngeal Carcinoma by Plasma Epstein-Barr Virus DNA Analysis in a Surveillance Program." *Cancer* 119, no. 10 (2013): 1838-44.

Chen, Wei, and Guo-Hua Hu. "Biomarkers for Enhancing the Radiosensitivity of Nasopharyngeal Carcinoma." *Cancer biology & medicine* 12, no. 1 (2015): 23.

Chou, Josephine, Yu-Ching Lin, Jae Kim, Liang You, Zhidong Xu, Biao He, and David M Jablons. "Nasopharyngeal Carcinoma—Review of the Molecular Mechanisms of Tumorigenesis." *Head & Neck: Journal for the Sciences and Specialties of the Head and Neck* 30, no. 7 (2008): 946-63.

Chu, Wing-Keung, Chih-Chin Hsu, Shiang-Fu Huang, Chia-Chi Hsu, and Shu-Er Chow. "Caspase 12 Degrades Iκbα Protein and Enhances Mmp-9 Expression in Human Nasopharyngeal Carcinoma Cell Invasion." *Oncotarget* 8, no. 20 (2017): 33515.

Cohen, Ezra, and Anne Lee. "Editorial for Special Issue on Nasopharynx Cancer." *Oral oncology* 50, no. 5 (2014): 325.

Decock, Julie, Sally Thirkettle, Laura Wagstaff, and Dylan R Edwards. "Matrix Metalloproteinases: Protective Roles in Cancer." *Journal of cellular and molecular medicine* 15, no. 6 (2011): 1254-65.

El-Badrawy, Mohamed K, Aida M Yousef, Dalia Shaalan, and Ayman Z Elsamanoudy. "Matrix Metalloproteinase-9 Expression in Lung Cancer Patients and Its Relation to Serum Mmp-9 Activity, Pathologic Type, and Prognosis." *Journal of bronchology & interventional pulmonology* 21, no. 4 (2014): 327-34.

Farina, Antonietta, and Andrew Mackay. "Gelatinase B/Mmp-9 in Tumour Pathogenesis and Progression." *Cancers* 6, no. 1 (2014): 240-96.

Friedberg, Marc H, Michael J Glantz, Mark S Klempner, Bernard F Cole, and George Perides. "Specific Matrix Metalloproteinase Profiles in the Cerebrospinal Fluid Correlated with the Presence of Malignant Astrocytomas, Brain Metastases, and Carcinomatous Meningitis."

Cancer: Interdisciplinary International Journal of the American Cancer Society 82, no. 5 (1998): 923-30.

Hsu, Chih-Chin, Shiang-Fu Huang, Jong-Shyan Wang, Wing-Keung Chu, Ju-En Nien, Wei-Shan Chen, and Shu-Er Chow. "Interplay of N-Cadherin and Matrix Metalloproteinase 9 Enhances Human Nasopharyngeal Carcinoma Cell Invasion." *BMC cancer* 16, no. 1 (2016): 800.

Huang, Hao. "Matrix Metalloproteinase-9 (Mmp-9) as a Cancer Biomarker and Mmp-9 Biosensors: Recent Advances." *Sensors* 18, no. 10 (2018): 3249.

Jeyakumar, Anita, Todd M Brickman, Alwin Jeyakumar, and Timothy Doerr. "Review of Nasopharyngeal Carcinoma." *Ear, nose & throat journal* 85, no. 3 (2006): 168-84.

Jia, Zan-Hui, Yan Jia, Feng-Jun Guo, Jun Chen, Xi-Wen Zhang, and Man-Hua Cui. "Phosphorylation of Stat3 at Tyr705 Regulates Mmp-9 Production in Epithelial Ovarian Cancer." *PloS one* 12, no. 8 (2017): e0183622.

Kondratiev, Svetlana, Douglas R Gnepp, Evgeny Yakirevich, Edmond Sabo, Donald J Annino, Elie Rebeiz, and Nora V Laver. "Expression and Prognostic Role of Mmp2, Mmp9, Mmp13, and Mmp14 Matrix Metalloproteinases in Sinonasal and Oral Malignant Melanomas." *Human pathology* 39, no. 3 (2008): 337-43.

Lam, Ka-On, Anne WM Lee, Cheuk-Wai Choi, Henry CK Sze, Anthony L Zietman, Kirsten I Hopkins, and Eduardo Rosenblatt. "Global Pattern of Nasopharyngeal Cancer: Correlation of Outcome with Access to Radiation Therapy." *International Journal of Radiation Oncology* Biology* Physics* 94, no. 5 (2016): 1106-12.

Lee, Sujin, Mei Zheng, Bumseok Kim, and Barry T Rouse. "Role of Matrix Metalloproteinase-9 in Angiogenesis Caused by Ocular Infection with Herpes Simplex Virus." *The Journal of clinical investigation* 110, no. 8 (2002): 1105-11.

Li, Zhi, Li-Juan Bian, Yang Li, Ying-Jie Liang, and Hui-Zhen Liang. "Expression of Protease-Activated Receptor-2 (Par-2) in Patients with Nasopharyngeal Carcinoma: Correlation with Clinicopathological

Features and Prognosis." *Pathology-Research and Practice* 205, no. 8 (2009): 542-50.

Liang, Shucai, and Lulin Chang. "Serum Matrix Metalloproteinase-9 Level as a Biomarker for Colorectal Cancer: A Diagnostic Meta-Analysis." *Biomarkers in medicine* 12, no. 4 (2018): 393-402.

Lin, C-W, S-W Tseng, S-F Yang, C-P Ko, C-H Lin, L-H Wei, M-H Chien, and Y-S Hsieh. "Role of Lipocalin 2 and Its Complex with Matrix Metalloproteinase-9 in Oral Cancer." *Oral diseases* 18, no. 8 (2012): 734-40.

Liu, Zhen, Lixia Li, Zhixiong Yang, Weiren Luo, Xin Li, Huiling Yang, Kaitai Yao, Bin Wu, and Weiyi Fang. "Increased Expression of Mmp9 Is Correlated with Poor Prognosis of Nasopharyngeal Carcinoma." *BMC cancer* 10, no. 1 (2010): 270.

Manabe, Shin-ichi, Zezong Gu, and Stuart A Lipton. "Activation of Matrix Metalloproteinase-9 Via Neuronal Nitric Oxide Synthase Contributes to Nmda-Induced Retinal Ganglion Cell Death." *Investigative ophthalmology & visual science* 46, no. 12 (2005): 4747-53.

Mehner, Christine, Alexandra Hockla, Erin Miller, Sophia Ran, Derek C Radisky, and Evette S Radisky. "Tumor Cell-Produced Matrix Metalloproteinase 9 (Mmp-9) Drives Malignant Progression and Metastasis of Basal-Like Triple Negative Breast Cancer." *Oncotarget* 5, no. 9 (2014): 2736.

Merdad, Adnan, Sajjad Karim, Hans-Juergen Schulten, Ashraf Dallol, Abdelbaset Buhmeida, Fatima Al-Thubaity, Mamdooh A Gari, et al. "Expression of Matrix Metalloproteinases (Mmps) in Primary Human Breast Cancer: Mmp-9 as a Potential Biomarker for Cancer Invasion and Metastasis." *Anticancer research* 34, no. 3 (2014): 1355-66.

Mira, Emilia, Rosa Ana Lacalle, José María Buesa, Gonzalo González de Buitrago, Sonia Jiménez-Baranda, Concepción Gómez-Moutón, Carlos Martínez-A, and Santos Mañes. "Secreted Mmp9 Promotes Angiogenesis More Efficiently Than Constitutive Active Mmp9 Bound to the Tumor Cell Surface." *J Cell Sci* 117, no. 9 (2004): 1847-57.

Murono, Shigeyuki, Tomokazu Yoshizaki, Hiroshi Sato, Hajime Takeshita, Mitsuru Furukawa, and Joseph S Pagano. "Aspirin Inhibits Tumor Cell

Invasiveness Induced by Epstein-Barr Virus Latent Membrane Protein 1 through Suppression of Matrix Metalloproteinase-9 Expression." *Cancer research* 60, no. 9 (2000): 2555-61.

NCCN. "Practice Guidelines in Oncology Head and Neck Cancers." *Head and Neck Cancer* (2010).

Pryczynicz, Anna, Katarzyna Guzińska-Ustymowicz, Violetta Dymicka-Piekarska, Jolanta Czyzewska, and Andrzej Kemona. "Expression of Matrix Metalloproteinase 9 in Pancreatic Ductal Carcinoma Is Associated with Tumor Metastasis Formation." *Folia histochemica et cytobiologica* 45, no. 1 (2007): 37-40.

Pua, KC, AS Khoo, YY Yap, SK Subramaniam, CA Ong, G Gopala Krishnan, and H Shahid. "Nasopharyngeal Carcinoma Database." *Med J Malaysia* 63, no. Suppl C (2008): 59-62.

Puspitasari, Dewi. *Gambaran Penderita Karsinoma Nasofaring Di Rsup H. Adam Malik Medan Tahun 2006-2010.* 2011. [*Overview of Nasopharyngeal Carcinoma Patients in Rsup H. Adam Malik Medan in 2006-2010.*]

Ren, Fanghui, Ruixue Tang, Xin Zhang, Wickramaarachchi Mihiranganee Madushi, Dianzhong Luo, Yiwu Dang, Zuyun Li, Kanglai Wei, and Gang Chen. "Overexpression of Mmp Family Members Functions as Prognostic Biomarker for Breast Cancer Patients: A Systematic Review and Meta-Analysis." *PloS one* 10, no. 8 (2015): e0135544.

Ruokolainen, Henni, Paavo Pääkkö, and Taina Turpeenniemi-Hujanen. "Expression of Matrix Metalloproteinase-9 in Head and Neck Squamous Cell Carcinoma: A Potential Marker for Prognosis." *Clinical Cancer Research* 10, no. 9 (2004): 3110-16.

Sakata, Koh-ichi, Masanori Someya, Mutsuko Omatsu, Hiroko Asanuma, Tadashi Hasegawa, Masato Hareyama, and Tetsuo Himi. "The Enhanced Expression of the Matrix Metalloproteinase 9 in Nasal Nk/T-Cell Lymphoma." *BMC cancer* 7, no. 1 (2007): 229.

Salehiniya, H, M Mohammadian, A Mohammadian-Hafshejani, and N Mahdavifar. "Nasopharyngeal Cancer in the World: Epidemiology, Incidence, Mortality and Risk Factors." *World Cancer Research Journal* 5, no. 1 (2018).

Stevenson, David, Chrystalla Charalambous, and Joanna B Wilson. "Epstein-Barr Virus Latent Membrane Protein 1 (Cao) up-Regulates Vegf and Tgfα Concomitant with Hyperlasia, with Subsequent up-Regulation of P16 and Mmp9." *Cancer research* 65, no. 19 (2005): 8826-35.

Sun, Wei, Dong-Bo Liu, Wen-Wen Li, Lin-Li Zhang, Guo-Xian Long, Jun-Feng Wang, Qi Mei, and Guo-Qing Hu. "Interleukin-6 Promotes the Migration and Invasion of Nasopharyngeal Carcinoma Cell Lines and Upregulates the Expression of Mmp-2 and Mmp-9." *International journal of oncology* 44, no. 5 (2014): 1551-60.

Tabuchi, Keiji, Masahiro Nakayama, Bungo Nishimura, Kentaro Hayashi, and Akira Hara. "Early Detection of Nasopharyngeal Carcinoma." *International journal of otolaryngology* 2011 (2011).

Tang, Jian-guo, Xuan Li, and Ping Chen. "Expression of Matrix Metalloproteinase-9 in Nasopharyngeal Carcinoma and Association with Epstein-Barr Virus Infection." *Journal of Zhejiang University-SCIENCE A* 5, no. 10 (2004): 1304-12.

Venugopal, Archana, and TN Uma Maheswari. "Expression of Matrix Metalloproteinase-9 in Oral Potentially Malignant Disorders: A Systematic Review." *Journal of oral and maxillofacial pathology: JOMFP* 20, no. 3 (2016): 474.

Vilen, Suvi-Tuuli, Tuula Salo, Timo Sorsa, and Pia Nyberg. "Fluctuating Roles of Matrix Metalloproteinase-9 in Oral Squamous Cell Carcinoma." *The Scientific World Journal* 2013 (2013).

Walker, Andrew, Rhett Frei, and Kathryn R Lawson. "The Cytoplasmic Domain of N-Cadherin Modulates Mmp-9 Induction in Oral Squamous Carcinoma Cells." *International journal of oncology* 45, no. 4 (2014): 1699-706.

Wang, Maoxin, Yilong Xu, Xianming Chen, Hui Chen, Hongxun Gong, and Shiyan Chen. "Prognostic Significance of Residual or Recurrent Lymph Nodes in the Neck for Patients with Nasopharyngeal Carcinoma after Radiotherapy." *Journal of cancer research and therapeutics* 12, no. 2 (2016): 909.

Weiwei, Yu, Liu Jinhui, Xiong Xiaoliang, Ai Yousheng, and Wang Huamin. "Expression of Mmp9 and Cd147 in Invasive Squamous Cell Carcinoma of the Uterine Cervix and Their Implication." *Pathology-Research and Practice* 205, no. 10 (2009): 709-15.

Wu, Qiu-Wan, Qing-Mo Yang, Yu-Fan Huang, Hong-Qiang She, Jing Liang, Qiao-Lu Yang, and Zhi-Ming Zhang. "Expression and Clinical Significance of Matrix Metalloproteinase-9 in Lymphatic Invasiveness and Metastasis of Breast Cancer." *PloS one* 9, no. 5 (2014): e97804.

Xiao, Guangli, Yabing Cao, Xibin Qiu, Weihua Wang, and Yufeng Wang. "Influence of Gender and Age on the Survival of Patients with Nasopharyngeal Carcinoma." *BMC cancer* 13, no. 1 (2013): 226.

Yang, Gui, Qiaoling Deng, Wei Fan, Zheng Zhang, Peipei Xu, Shihui Tang, Ping Wang, and Mingxia Yu. "Cyclooxygenase-2 Expression Is Positively Associated with Lymph Node Metastasis in Nasopharyngeal Carcinoma." *PloS one* 12, no. 3 (2017): e0173641.

Yang, Jing, Xing Lv, Jinna Chen, Changqing Xie, Weixiong Xia, Chen Jiang, Tingting Zeng, et al. "Ccl2-Ccr2 Axis Promotes Metastasis of Nasopharyngeal Carcinoma by Activating Erk1/2-Mmp2/9 Pathway." *Oncotarget* 7, no. 13 (2016): 15632.

Yilmaz, Mahmut, and Gerhard Christofori. "Mechanisms of Motility in Metastasizing Cells." *Molecular cancer research* 8, no. 5 (2010): 629-42.

Yip, Timothy TC, Roger KC Ngan, Alvin HW Fong, and Stephen CK Law. "Application of Circulating Plasma/Serum Ebv DNA in the Clinical Management of Nasopharyngeal Carcinoma." *Oral oncology* 50, no. 6 (2014): 527-38.

Yoshizaki, Tomokazu, Hiroshi Sato, Mitsuru Furukawa, and Joseph S Pagano. "The Expression of Matrix Metalloproteinase 9 Is Enhanced by Epstein–Barr Virus Latent Membrane Protein 1." *Proceedings of the National Academy of Sciences* 95, no. 7 (1998): 3621-26.

Yu, Qin, and Ivan Stamenkovic. "Cell Surface-Localized Matrix Metalloproteinase-9 Proteolytically Activates Tgf-B and Promotes Tumor Invasion and Angiogenesis." *Genes & development* 14, no. 2 (2000): 163-76.

Zergoun, Ahmed-Amine, Abderezak Zebboudj, Sarah Leila Sellam, Nora Kariche, Djamel Djennaoui, Samir Ouraghi, Esma Kerboua, et al. "Il-6/Nos2 Inflammatory Signals Regulate Mmp-9 and Mmp-2 Activity and Disease Outcome in Nasopharyngeal Carcinoma Patients." *Tumor Biology* 37, no. 3 (2016): 3505-14.

Zhang, Caiyun, Chao Li, Minhui Zhu, Qingzhou Zhang, Zhenghua Xie, Gang Niu, Xicheng Song, et al. "Meta-Analysis of Mmp2, Mmp3, and Mmp9 Promoter Polymorphisms and Head and Neck Cancer Risk." *PloS one* 8, no. 4 (2013): e62023.

Zhou, Dong-Ni, Yan-Fei Deng, Rong-Hua Li, Ping Yin, and Chun-Sheng Ye. "Concurrent Alterations of Rage, Reck, and Mmp9 Protein Expression Are Relevant to Epstein-Barr Virus Infection, Metastasis, and Survival in Nasopharyngeal Carcinoma." *International journal of clinical and experimental pathology* 7, no. 6 (2014): 3245.

BIOGRAPHICAL SKETCH

Farhat Farhat

Affiliation: Department of Otorhinolaryngology Head and Neck Surgery, Faculty of Medicine, Universitas Sumatera Utara, Medan, Indonesia.

Education: Basic medical in Medical Faculty of Sumatera Utara University, Indonesia and graduated as general practitioner in 1997. Become a Otorhinolaryngologist in 2004. Appointed as a Consultant of Head and Neck Oncology in 2011. Graduated as Master in Medical Clinician in 2012, and become a Doctor of Otorhinolaryngology in 2014.

Research and Professional Experience:

Consultant of Head and Neck Oncology	Adam Malik General Hospital Medan
Researcher	Center of Excellence Nasopharyngeal Carcinoma of Universitas Sumatera Utara
Lecturer	Medical Faculty of Universitas Sumatera Utara

Professional Appointments:

Chairman	PERHATI-KL of North Sumatera
General Secretary	Indonesian Cancer Foundation of North Sumatera
Secretary	Partnership and International Relationship of Indonesian Medical Association (IDI) of Medan
Lecturer	Medical Faculty of Universitas Sumatera Utara
Instructor	Basic Surgical Skill Course
University Secretary	Universitas Sumatera Utara

Honors: Dr. dr, M.Ked(ORL-HNS), Sp.T.H.T.K.L(K)

Publications from the Last 3 Years:

Farhat, Farhat. (2017). The Increased Expressions of Type IV Collagen in Cochlear Fibroblasts of Diabetic Rat Models Caused by Curcumin Therapy. *Journal of Applied Pharmaceutical Science* Vol 7.

Farhat, Farhat. (2017). Profile of Sinonasal Malignant Tumor Patients in Adam Malik General Hospital Medan Indonesia. *Bali Medical Journal* Vol.7, Number 1: 137-140.

Farhat, Farhat. (2017). The Effectiveness of Frenotomy in the Treatment of Ankyloglossia; A Case Report from Adam Malik General Hospital Medan-Indonesia. *Bali Medical Journal* Vol.7, Number 1: 192-194.

Farhat, Farhat. (2018). The Antiapoptotic Effect of Curcumin in the Fibroblast of the Cochlea in an Ototoxic Rat Model. *Iranian journal of otorhinolaryngology* 30 (100), 247.

Farhat, Farhat. (2018). An Uncommon Occurrence of Pleomorphic Adenoma in the Submandibular Salivary Gland: A Case Report. *Open Access Macedonian Journal of Medical Sciences* 6 (6), 1101-1103.

Farhat, Farhat. (2018). P38 mitogen-activated protein kinase (p38 MAPK) overexpression in clinical staging of nasopharyngeal carcinoma. *IOP Conference Series: Earth and Environmental Science* 125 (1), 012129.

Farhat, Farhat. (2018). Evaluation of matrix metalloproteinase-9 expressions in nasopharyngeal carcinoma patients. *IOP Conference Series: Earth and Environmental Science* 125 (1), 012130.

Farhat, Farhat. (2018). Profile of sinonasal malignant tumor patients in Adam Malik General Hospital Medan Indonesia. *Bali Medical Journal* 7 (1), 137-140.

Farhat, Farhat. (2018). The effectiveness of frenotomy in the treatment of ankyloglossia: A case report from Adam Malik General Hospital Medan-Indonesia. *Bali Medical Journal* 7 (1), 192-194.

In: A Closer Look at Metalloproteinases ISBN: 978-1-53616-517-3
Editor: Lena Goodwin

Chapter 3

MATRIX METALLOPROTEINASES IN CARDIOVASCULAR DISEASE: CURRENT UPDATES AND NEW INSIGHTS

***M. Beutline Malgija*[*], PhD**
and C. Joyce Priyakumari, PhD
Bioinformatics Centre of BTISnet,
Madras Christian College, Chennai, Tamilnadu, India

ABSTRACT

Matrix metalloproteinases (MMPs), also called Martixins, functions in the extracellular environment of the cells and degrade both matrix and non-matrix proteins. They are a large family of zinc-endopeptidases which play vital roles in multiple physiological and pathological processes. Maintenance of the structural integrity of the cardiac extracellular matrix is important for proper functioning of the heart. MMPs were reported to cause changes in the structural framework of the extracellular matrix by stimulation of growth factors and inflammatory mediators, thereby causing abnormality in cardiac remodeling and inflammatory response in cardiovascular diseases. Since most of the MMPs are reported to have their role in pathological shift causing various diseases, they are considered as

[*] Corresponding Author's E-mail: beutline.bioinfo@gmail.com.

prominent therapeutic targets for preventive medicine. This chapter brings in the members of MMP family, their structural organization, function and discusses their role in cardiovascular disease.

Keywords: matrixins, matrix metalloproteinases, extracellular matrix, cardiovascular diseases, atherosclerosis, myocardial infarction, heart failure, myocarditis

INTRODUCTION

Matrix metalloproteinases (MMPs), also called matrixins are zinc-dependent protein and peptide hydrolases capable of degrading most of the extracellular matrix (ECM) components (Vihinen and Kahari, 2002) such as proteoglycans, insoluble collagen fibers and soluble ECM proteins (fibronectin, lamininetc), involved in various physiological and pathological conditions. They are synthesized as zymogens with a signal peptide which leads them to the secretory pathway. These enzymes can then be secreted from the cell or anchored to the plasma membrane, thereby confining their catalytic activity to the extracellular space or to the cell surface, respectively. More than 25 types of MMPs have been identified in humans and many are reported to have their role in inflammation, arthritis, cancer and cardiovascular diseases. In cardiovascular diseases, they play a central role in atherosclerosis, plaque formation, platelet aggregation, acute coronary syndrome, restenosis, aortic aneurysms and peripheral vascular disease (Papazafiropoulou and Tentolouris, 2009). Specifically, MMP-1, -2, -3, -8, -9, -12, -13, -14, -28 has been identified to be involved in cardiac remodeling. The role of higher MMPs in the cardiovascular system is less well explored (Pytliak et al., 2017). Detailed structural knowledge is essential for the design of drugs, as the key therapeutic strategy to combat the deregulation of MMPs target their catalytic domains. This chapter reviews the structural and functional aspects of various MMP types and their biological role in normal cardiac activity and the abnormalities associated with them, in particular their involvement in CVDs.

MMP CLASSIFICATION

The MMP family shares a similar basic domain structure and can be divided into different groups (Figure 1) based on structure and *in vitro* substrate specificity for various ECM components (Table 1).

Collagenases

Collagenases is composed of three enzymes, MMP-1, MMP-8 and MMP-13 (also known as collagenases-1, 2, and 3, respectively) whose name reflects their ability to cleave the collagen triple helix into characteristic 3/4 and 1/4 fragments. They are also able to proteolytically process other ECM proteins, as well as a number of bioactive molecules such as interleukin-8 (IL-8) (Tester et al., 2007), pro-tumor necrosis factor (TNF)-α, protease-activated receptor-1 (Boire et al., 2005), and several insulin-like growth factor binding proteins (IGFBPs) (Overall, 2002). Removal of the hemopexin domain transforms these MMPs into enzymes uncapable to degrade native collagen, signifies that the collaboration between the catalytic and hemopexin domains is essential to carry out their collagenolytic activity (Arnold et al., 2011).

Stromelysins

Stromelysins including MMP-3 (Stromelysin-1) and MMP-10 (Stromelysin-2) are capable of degrading many different ECM components, but incapable to cleave native collagen. They digest a number of ECM molecules and participate in proMMP activation. They also partake in proMMP activation by their ability to remove the propeptide domain of the three procollagenases (Barksby et al., 2006) and proMMP-9 (Geurts et al., 2008), generating fully activated form of the enzymes. Stromelysins are expressed by both fibroblast and epithelial cells, and are secreted to the extracellular space where they play vital roles in various biological processes. Another MMP called stromelysin-3 (MMP-11) shares some structural characteristics with stromelysins, but due to the presence of additional features later classified into the category of furin-activatable MMPs (Fanjul-Fernández et al., 2010).

Matrilysins

Matrilysins including MMP-7 (Matrilysin-1) and MMP-26 (Matrilysin-2), exhibit the simplest domain arrangement of all MMPs since they lack the carboxy-terminal hemopexin domain (Uria and Lopez-Otin, 2000). Matrilysins play essential roles in the degradation of ECM proteins like type IV collagen, laminin and entactin (Overall, 2002). MMP-7 catalyzes the ectodomain shedding of several cell surface molecules like E-cadherin (McGuire et al., 2003), Fas ligand (Wang et al., 2006) and syndecan-1 (Li et al., 2002), whereas MMP-26 has been found to be an activator of proMMP-9 under pathological conditions (Zhao et al., 2004). It digests several ECM molecules, and unlike most other MMPs, it is largely stored intracellularly (Marchenko et al., 2004).

Gelatinases

MMP-2 (gelatinase-A) and MMP-9 (gelatinase-B), also known as Gelatinases have an additional fibronectin domain located inside the catalytic domain, which permits the binding and processing of denatured collagen or gelatin (Gomis-Ruth, 2009), suggesting their key role in the remodeling of ECM. They play a major role in degradation of collagen chains and in cleavage of ECM components. Both gelatinases have been related with inflammatory disorders and vascular alterations such as atherosclerosis, aortic aneurysm and myocardial infarction (Hu et al., 2007). They also play a fundamental role in regulating vascular smooth muscle cell migration and proliferation (Johnson, 2007).

Table 1. MMPs and their substrates

Group	Name	MMP	Collagenous substrates
Collagenases	Interstitial collagenase	MMP-1	Collagen types I, II, III, V, VII, VIII, X and gelatin, Aggrecan, elastin, serpine, versican, perlecan, proteoglycan link protein, tenascin C
	Neutrophil collagenase	MMP-8	Collagen types I, II, III, IV, V, IX, X, XI and gelatin, Aggrecan, laminin
	Collagenase-3	MMP-13	Collagen I, gelatin, Aggrecan, fibronectin, laminin,
	Collagenase-4	MMP-18	perlecan, tenascin
Gelatinases	Gelatinase A	MMP-2	Collagen types I, IV, V, VII, X, XI, XIV and gelatin, Aggrecan, elastin, fibronectin, laminin, proteoglycan link protein, versican
	Gelatinase B	MMP-9	Collagen types V, VI, VII, X, XIV and gelatin, Elastin, fibronectin, vitronectin, laminin, proteoglycan link protein, versican
Stromelysins	Stromelysin 1	MMP-3	Collagen types II, IV, IX, X and gelatin, Aggrecan, casein, elastin, fibronectin, laminin, perlecan, proteoglycan link protein, versican, MMP-2/TIMP-2
	Stromelysin 2	MMP-10	Collagen II, IV, V and gelatin, Fibronectin, laminin
	Stromelysin 3	MMP-11	Fibronectin, laminin, gelatin, aggrecan
Matrilysins	Matrilysin 1	MMP-7	Collagen types I, II, III, V, VI and X Collagen IV, gelatin, Fibronectin, vitronectin, laminin, gelatin, aggrecan
	Matrilysin 2	MMP-26	fibrin, Collagen, type IV and gelatin, fibrinogen, casein
Elastase	Metalloelastase	MMP-12	Collagen IV, gelatin, Elastin, fibronectin, laminin, vitronectin, proteoglycan
Furin-activated MMPs		MMP-11	Fibronectin, laminin, gelatin, aggrecan
		MMP-21	
		MMP-23	Gelatin, fibronectin
Membrane Type	*Transmembrane:*		
	MT1-MMP	MMP-14	Pro-MMP2, procollagenase 3
	MT2-MMP	MMP-15	Pro-MMP2
	MT3-MMP	MMP-16	Collagen types I and III, gelatin, aggrecan, fibronectin, laminin, vitronectin, proteoglycan
	MT5-MMP	MMP-24	Gelatin, Fibrin, fibronectin
	GPI- anchored:		
	MT4-MMP	MMP-17	Gelatin, Fibrin, fibronectin
	MT6-MMP	MMP-25	Collagen IV, gelatin, casein, fibrinogen, fibronectin
	Type II	MMP-23	Gelatin, fibronectin
Other MMPs	Enamelysin	MMP-19	Collagen IV, gelatin
		MMP-20	Amelogenin, aggrecan
		MMP-22	
		MMP-27	
		MMP-28	

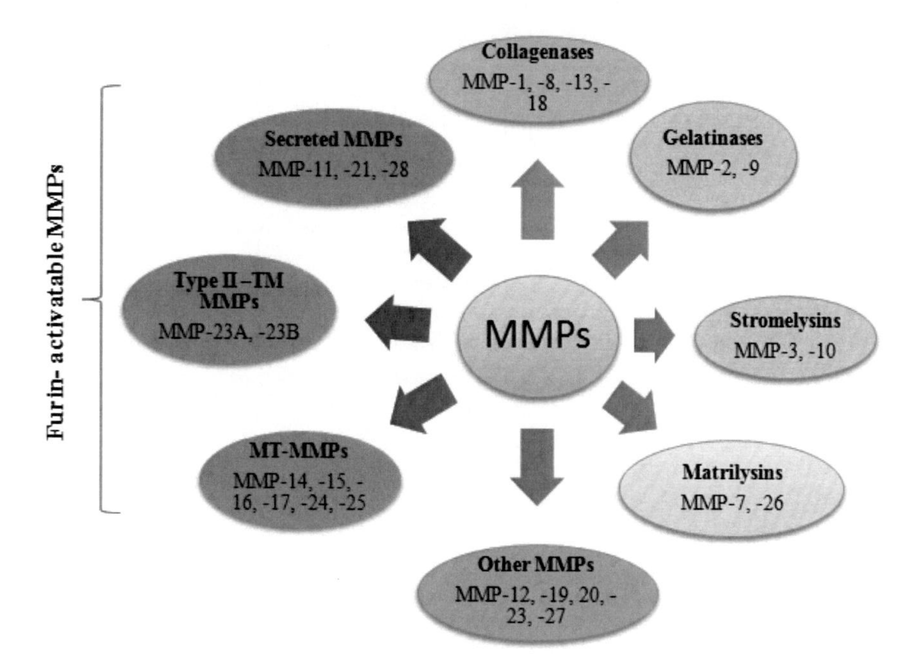

Figure 1. Classification of MMPs. The three MMP groups namely secreted, Type-II Transmembrane (TM) and membrane type (MT) are furin activatable MMPs.

FurinActivatable MMPs

MMPs belonging to this category hold a furin recognition motif inserted between the propeptide and the catalytic domain. These include three secreted MMPs (MMP-11, -21, and -28), six membrane-type-MMPs (MMP-14, -15, -16, -17, -24 and -25) and two unusual type II transmembrane MMPs (MMP-23A and MMP-23B).

Secreted MMPs

This subgroup comprises MMP-11, -21 and -28. Unlike the other secreted MMPs, these furin-activatable enzymes are processed intracellularly by furin-like proteases as active forms. MMP-11, also known as stromelysin-3, is expressed during embryogenesis, tissue involution and wound healing (Rio, 2005).

Membrane-Type MMPs

This group includes four type I transmembrane proteins (MMP-14, -15, -16, and -24) and two glycosylphosphatidylinositol-anchored proteins (MMP-17 and -25). They possess a furin recognition sequence at the C-terminus of the peptide, therefore activated intracellularly and likely to be expressed on the cell surface.

Type II Transmembrane MMPs

These comprises MMP-23A and MMP-23B, having identical amino acid sequence, but are encoded by distinct genes. They have unique C-terminal cysteine-rich immunoglobulin-like domains instead of a hemopexin domain (Nagase et al., 2006). Expression analysis showed the predominance of MMP-23 in ovary, testis and prostate, suggesting the possibility of their specialized role in reproductive processes (Velasco et al., 1999).

Other MMPs

MMP-12, also known as macrophage metalloelastase was first identified as a secretory protein from mouse peritoneal macrophage through its elastase activity (Were and Gordon, 1975). It is a key player in tissue remodeling associated with many pathological conditions such as chronic inflammation and fibrosis (Klein and Bischoff, 2011). Epilysin (MMP-28) is expressed in many tissues such as placenta, lung, heart, gastrointestinal tract and testis (Saarialho-Kere et al., 2002). It is also highly expressed in cartilage from patients with osteoarthritis and rheumatoid arthritis (Kevorkian et al., 2004; Kere, 2004).

STRUCTURAL ORGANIZATION

MMPs generally contains three homologous protein domains: i) the signal peptide, which is a domain that targets the enzyme for secretion ii) the propeptide domain, which is removed on activation of the enzyme and iii) the catalytic domain containing the zinc binding region and is responsible for the proteolytic activity of the enzyme.

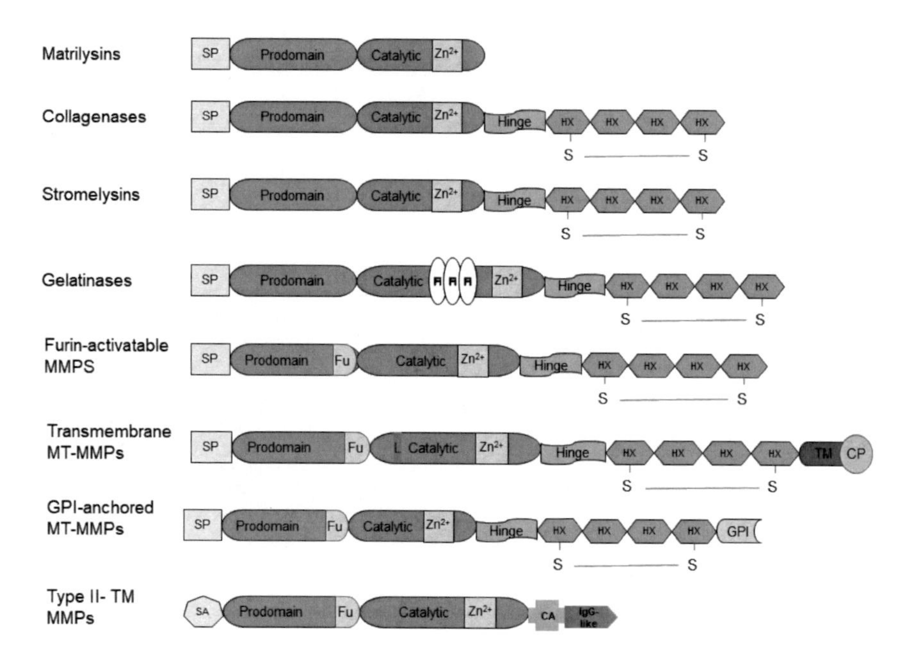

Figure 2. Differences in domain organization between various classes of MMPs. Matrilysins lack the hinge region and hemopexin domains. Collagenases and stromelysins share the same structural domains. In Gelatinases, the catalytic domain contains the gelatin-binding domain, which is homologous with the collagen-binding domain of fibronectin. The hemopexin domain has been reported to play a functional role in substrate binding and interactions with tissue inhibitors of MMPs. TM MT-MMPs contains a transmembrane domain and a short cytoplasmic domain, whereas the glycosylphosphatidyl-inositol (GPI)-anchored type contains a short hydrophobic sequence at the end, that functions as a GPI-anchoring signal peptide. TM MT-MMPs harbor an 8-aminoacid loop in the catalytic domain named the MT-loop, which is unique to TM-type MT-MMPs amongst all the MMPs.

Differences in the structural domain of the various MMP families are depicted in Figure 2. Secondary structural arrangement of stromelysins showed more number of β-sheets as compared to α-helix based on our previous study on structural comparison of the predicted three dimensional models of stromelysins (Malgija et al., 2018). As most of MMPs lack a perfect three dimensional crystallographic structures, *in silico* prediction of the remaining structures will provide more clues on their structure and function.

CARDIOVASCULAR DISEASES (CVDS)

CVDs refer to a general term for conditions affecting heart and blood vessels. This includes coronary artery diseases such as angina and myocardial infarction (MI), hypertensive heart disease, cardiomyopathy, stroke, rheumatic heart disease, congenital heart disease, valvular heart disease, aortic aneurisms, carditis, pherepheral artery disease and more. Abnormalities in normal biological processes associated with heart such as ECM remodeling, ventricular remodeling and inflammatory processes can contribute to CVD. MMPs and their role in contribution of such abnormalities, leading to CVDs is discussed below.

ANOMALIES IN BIOLOGICAL PROCESSES ASSOCIATED WITH CARDIOVASCULAR DISEASE

Cardiac Remodeling

Cardiac remodeling, which refers to the structural and functional changes in the heart, mainly includes changes in cardiac myocytes and in the ECM. The myocardial ECM is a complex network which determines the structural integrity of the heart. ECM signifies an important cardiac element that adapts to coordinate the functional necessities of the myocardium. It contains a wide variety of structural proteins, such as collagens, proteoglycans and glycosaminoglycans, and serves as a repository for biologically active molecules. Since myocardial collagens maintain the structural integrity of adjoining myocytes and provide the means by which myocyte shortening is translated into cardiac pump function, changes in the ECM result in loss of normal structural and functional myocardium (Phatharajaree et al., 2007). MMPs release ECM fragments called matricryptins or matrikines that plays an important role in development of heart failure and post myocardial infarction (Lindsey et al., 2015; Yabluchanskiy et al., 2013a) and multiple studies proposed matricryptins as potential therapeutic targets for heart failure patients. They also regulate the remodeling process by facilitating ECM turnover and inflammatory signaling (DeLeon-Pennell et al., 2017). MMPs regulate remodeling process in the myocardium by facilitating ECM turnover and inflammatory signals. Adverse cardiac remodeling that involves elevated ECM turnover contributes to high morbidity and mortality in myocardial infarction patients (Cohn et al., 2000). Elevated serum MMP-7 levels are linked with left ventricular (LV) structural remodeling in LV hypertrophy (Zile et al., 2011). Serum levels of MMP-8 are a significant predictor of LV remodeling, cardiac rupture and development of heart failure after MI (Iyer et al., 2014; Fertin et al., 2013). MMP-9 regulates tissue remodeling by directly degrading ECM and activating cytokines and chemokines (Yabluchanskiy et al., 2013b) and is also one of the potential markers for cardiac modeling, evident from both animal models and clinical studies (Halade et al., 2013). Squire et al., (2004) revealed the correlation of increased MMP-9 with larger LV volumes and greater LV dysfunction following MI. Increased MMP-14 in both plasma and the LV infarct post-MI associates with extensive LV remodeling including significant cardiac fibrosis, reduced LV function and lower survival (Wilson et al., 2003; Lindsey and Zamilpa, 2012).

Ventricular Dysfunction

Elevated MMP levels strongly correlate with LV dysfunction in CVD patients. Studies suggested MMP-9 as a novel prognostic biomarker for the development of LV dysfunction and late survival in patients with CVD (Kelly et al., 2007; Blankenberg et al., 2003). ECM proteins like MMP-16, COMP, ELN, collagen genes (COL1A1, COL1A2, COL3A1, COL5A1, COL5A2, COL14A1 and COL16A1) were found to be highly expressed (Malgija and Shanmughavel, 2015; Malgija et al., 2018) in our previous study on hypertrophic and dilated cardiomyopathy. Increased presence of ECM proteins in the myocardium causes alteration in ventricular function, leading to systolic and diastolic dysfunction (Klein et al., 2005). It has been reported that MMP-2 increases with the severity of LV dysfunction in HCM patients (Noji et al., 2004).

Inflammatory Response

The maintenance of the physiological myocardial matrix turnover involves a highly regulated interaction between cardiac and noncardiac cells, in response to the release of inflammatory mediators and components of the matrix degrading system (Rutschow et al., 2005). Proinflammatory cytokines like TNF-α and IL-1β contribute to depression of LV function and cardiomyocyte loss by apoptosis. MMP-12 can process pro-TNFα into mature TNFα, indicating its potential to amplify TNFα-driven inflammation (Consili et al., 2013). It has broad substrate specificity for various ECM components and plays a key role in tissue remodeling linked with many pathological conditions such as chronic inflammation and fibrosis (Klein and Bischoff, 2011). Following MI, MMPs facilitate ECM degradation and recruit inflammatory cells for removal of necrotic cardiomyocytes. At the outset, the increase in expression of pro-inflammatory cytokines results in robust MMP activation, however, long-term stimulation increases tissue inhibitor of metalloproteinase (TIMP) levels. This ultimately leads to a decrease in the MMP/TIMP ratio and results in ongoing long-term remodeling (Wilson et al., 2002).

Fibrosis

Myocardium comprises a number of cell types namely cardiofibroblasts (CFBs), cardiomyocytes, endothelial cells and smooth muscle cells. CFBs are unique among other cell types in myocardium as they lack a basement membrane.

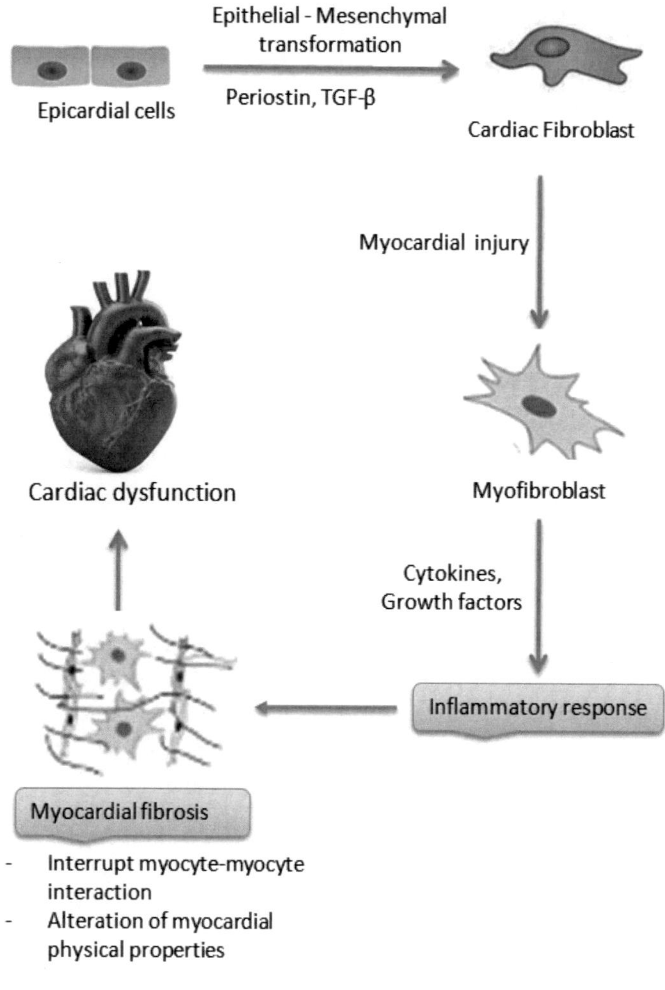

Figure 3: Formation of Myofibroblast and their role in fibrosis

In response to appropriate stimuli, especially during myocardial injury, CFBs can differentiate into myofibroblasts (myoFBs), which has greater ability to produce ECM proteins (Petrov et al., 2002). Differentiation from fibroblast to myofibroblast is induced by transforming growth factor beta (TGF-β), cytokines, the ECM and other growth factors (Tamaoki et al., 2005; Walker et al., 2004). Myofibroblasts have been verified to play a major role in reparative fibrosis in the infracted heart and with hypertrophic fibrotic scars (Calderone et al., 2006). They are absent in healthy myocardium and only appear after myocardial injury (Baum et al., 2011). Pressure overload causes a biomechanical stress on the ventricles and can trigger cardiac hypertrophy and fibrosis. The excessive stress is transmitted to ECM and cell-ECM connection, leading to adverse remodeling of the ECM, further activates the intracellular signaling pathways leading to cardiac hypertrophy, fibrosis and cell death.

MMPs in Cardiovascular Diseases

MMPs and their natural inhibitors play a major role in the remodeling of ECM in both normal and pathological conditions. In addition, MMPs have an important role in cardiovascular diseases, including MI, atherosclerosis, cardiomyopathy and heart failure.

Myocardial Infarction

MI, also known as heart attack, occurs when blood flow to a part of the heart declines or stops, causing damage to the heart muscle. Elevation in MMP levels after MI correlates with LV dysfunction in heart failure patients. MMP-7 activity is associated to higher risk for major adverse cardiac events, including decreases post-MI survival and increased hospitalization for congestive heart failure (Zile et al., 2011; Chiao et al., 2010). Clinical evidence reveals that genetic polymorphisms can also contribute to MMP protein levels, thus influencing cardiovascular outcomes (Spinale 2004). Due to their ability to regulate MMP-3 activity, MMP-3 polymorphisms have been implicated as regulators of MI prevalence and heart failure outcomes (Wang et al., 2011). A few studies reported the association of MMP-9 -1562 C/T polymorphism (Rodriguez-Perez et al., 2016; Shi, 2014) and MMP-12 -82 A/G polymorphism (Perez-Hernandez et al., 2012) with increased MI incidence.

Heart Failure

Heart failure (HF) is the common problem that resides the final end stage in most of the cardiovascular diseases. In patients with acute heart failure (AHF), production of MMPs is affected by several mechanisms including changes in hemodynamic conditions and neurohormonal and inflammatory factors (Pytliak et al., 2017). Biolo et al., 2010 reported the increase in expression of some markers associated with ECM turnover including MMP-2, TIMP-1 and procollagen type III N-terminal peptide in AHF syndrome. The increase in ECM turnover may be associated with an acceleration of pathological remodeling.

Cardiomyopathy

Cardiomyopathy is a disease of the heart muscle that makes difficulty to the heart in pumping blood to the rest of the body. MMPs are important in the process of LV remodeling in cardiomyopathy and is implicated in both ventricular hypertrophy and dilation (Spinale, 2002). Although MMPs are associated with LV remodeling in patients with hypertrophic cardiomyopathy (HCM), the impact of their plasma MMP levels are vague. Kitaoka et al., (2011) reported the association of MMP2 and reduced systolic function. MMP-9 is associated with small LV size and degree of LV hypertrophy, suggesting the importance of MMPs in the process of LV remodeling in HCM. Studies also demonstrated the elevated expression of MMP-9 in dilated cardiomyopathy (Morine et al., 2016; Nepomnyashchikh et al., 2015). MMP-2, MMP-9, TIMP-1, and the MMP/TIMP ratios can be promising predictors of dilated cardiomyopathy and can be used for assessment of treatment effectiveness in these conditions (Antonov et al., 2012). Collagen turn-over is enhanced in HCM, maintaining collagen I synthesis through degradation (Lombardi et al., 2003). An increase in collagen content was observed in young HCM patients, who died suddenly. These features are associated with changes in MMP expression, mainly high levels of MMP-2 and MMP-9 (Noji et al., 2004; Lombardi et al., 2003). Effectively, we found the upregulation of MMP-16 in HCM (Malgija and Shanmughavel, 2015).

Atherosclerosis

The thrombogenic core of advanced atherosclerotic plagues is composed of extracellular lipids, cell debris, and lipid-laden macrophages. In stable plagues, this core is protected from the circulating blood by a fibrous cap, which is rich in type I, II and III fibrillar collagens. It has been reported that proteolysis of fibrillar collagens within the protective fibrous plague precipitates plague destabilization (Libby, 2013). Several studies indicated the influence of atherosclerotic lesion formation.MMP-10 has reported to localize in macrophage-rich regions and endothelial cells within atherosclerotic plagues (Montero et al., 2006). Recent findings have proposed that MMP-10 regulates macrophage migration and invasion (Murray et al., 2013), and induction by thrombin may exert a fibrinolytic action (Orbe et al., 2009). Consequently, it is suggested that MMP-10 is pro-atherogenic, nevertheless the direct role of MMP-10 in atherosclerosis has not yet been evaluated. Schonbeck et al., 1999 reported the upregulation of MMP-11 expression in endothelial cells, smooth muscle cells and macrophages within carotid and aortic plagues. Increased MMP-2 expression has been observed in patients with carotid atherosclerotic plague complications (Alvarez et al., 2004). Gelatinases, including MMP-2 and MMP-9 are needed for the proliferation and migration of vascular smooth muscle cells and probably play a prominent role in the development and maintenance of the fibrous cap (Johnson, 2017). Elevated MMP-9 levels are an independent risk factor of cardiovascular morality in patients with coronary artery disease (Blankenberg et al., 2003). Elevated plasma levels of MMP-7 have been related with symptomatic carotid and coronary atherosclerosis (Abbas et al., 2014; Nilsson et al., 2006). MMP12 regulates fibrosis in several models of tissue injury and has a protective effect on corneal fibrosis during wound repair through regulation of immune cell infiltration and angiogenesis (Chan et al., 2013).

Myocarditis

Myocarditis, an inflammatory disorder, which is mostly caused by viral infection, is related with acute LV dysfunction accompanied by myocardial inflammatory cell infiltration and increased release of proinflammatory cytokines. Acute LV dysfunction in experimental induced acute myocarditis is associated with induction of proinflammatory cytokines and an imbalance of the MMPs and TIMPS system (Li et al., 2002). In myocarditis, the main disruption of ECM in early phase is caused by qualitative changes in the collagen network (Li et al., 2002). Increased levels of MMPs, decreased levels of TIMPs and the activated plasmin system in the acute phase of myocarditis may lead to an imbalance in ECM. This ultimately reduces the matrix integrity and disrupts the three dimensional collagen networks by cleaving the cross-links between the collagen molecules, which then leads to LV dysfunction and dilation (Rutschow et al., 2005).

Conclusion and Future Aspects

Tremendous knowledge has been accumulated to show that matrixins play numerous roles in both biological and pathological processes. Biochemical studies and the available 3D structures of MMPs have provided the molecular basis of our understanding on the function of these multi-domain proteinases function and their interaction with ECM molecules and inhibitors. Structural studies also provide clues to manipulate their enzymatic activities. A large number of MMP inhibitors have also been designed and synthesized based on these studies and some were clinically tested, but they showed little efficacy so far. The failure of these clinical studies may be due to the limited knowledge on the function of MMPs in normal and pathological conditions. As metal binding groups modulates the selective inhibition of MMPs, more studies are required to assess their selectivity, thereby making a possibility to reduce such limitations.

The myocardial ECM is a highly complex network, constantly undergoing a remodeling process, the balance of which establishes structural integrity of the heart. The imbalance of the system with changes in expression of MMPs and plasminogen activators and the reduced expression of their inhibitors TIMPs, leads to collagen turnover with an impairment of LV function. Therefore, the regulation of MMP/TIMP system is important in prevention of their pathologic role. Further investigations are necessary to find the best target to control this complex system causing dynamic balance between collagen accumulation and degradation. Moreover, identifying the diverse and novel roles of ECM components, in particular the distinct physical, chemical, and mechanical properties, in various cellular and developmental processes are essential.

REFERENCES

Abbas, A., Aukrust, P., Russell, D., Krohg-Sørensen, K., Almas, T., Bundgaard, D., & Holm, S. (2014). Matrix metalloproteinase 7 is associated with symptomatic lesions and adverse events in patients with carotid atherosclerosis. *PloS one*, *9*(1), e84935.

Alvarez, B., Ruiz, C., Chacón, P., Alvarez-Sabin, J., &Matas, M. (2004). Serum values of metalloproteinase-2 and metalloproteinase-9 as related to unstable plaque and inflammatory cells in patients with greater than 70% carotid artery stenosis. *Journal of vascular surgery*, *40*(3), 469-475.

Antonov, I. B., Kozlov, K. L., Pal'tseva, E. M., Polyakova, O. V., & Lin'kova, N. S. (2018). Matrix Metalloproteinases MMP-1 and MMP-9 and Their Inhibitor TIMP-1 as Markers of Dilated Cardiomyopathy in Patients of Different Age. *Bulletin of experimental biology and medicine*, *164*(4), 550-553.

Arnold, L. H., Butt, L. E., Prior, S. H., Read, C. M., Fields, G. B., & Pickford, A. R. (2011). The interface between catalytic and hemopexin domains in matrix metalloproteinase-1 conceals a collagen binding exosite. *Journal of Biological Chemistry*, *286*(52), 45073-45082.

Barksby, H. E., Milner, J. M., Patterson, A. M., Peake, N. J., Hui, W., Robson, T., & Rowan, A. D. (2006). Matrix metalloproteinase 10 promotion of collagenolysis via procollagenase activation: implications for cartilage degradation in arthritis. *Arthritis & Rheumatism: Official Journal of the American College of Rheumatology*, *54*(10), 3244-3253.

Baum, J., & Duffy, H. S. (2011). Fibroblasts and myofibroblasts: what are we talking about?. *Journal of cardiovascular pharmacology*, *57*(4), 376.

Biolo, A., Fisch, M., Balog, J., Chao, T., Schulze, P. C., Ooi, H., ...& Colucci, W. S. (2010). Episodes of acute heart failure syndrome are associated with increased levels of troponin and extracellular matrix markers. *Circulation: Heart Failure*, *3*(1), 44-50.

Blankenberg, S., Rupprecht, H. J., Poirier, O., Bickel, C., Smieja, M., Hafner, G., ...& Tiret, L. (2003). Plasma concentrations and genetic variation of matrix metalloproteinase 9 and prognosis of patients with cardiovascular disease. *Circulation*, *107*(12), 1579-1585.

Boire, A., Covic, L., Agarwal, A., Jacques, S., Sherifi, S., & Kuliopulos, A. (2005). PAR1 is a matrix metalloprotease-1 receptor that promotes invasion and tumorigenesis of breast cancer cells. *Cell*, *120*(3), 303-313.

Calderone, A., Bel-Hadj, S., Drapeau, J., El-Helou, V., Gosselin, H., Clement, R., & Villeneuve, L. (2006). Scar myofibroblasts of the infarcted rat heart express natriuretic peptides. *Journal of cellular physiology*, *207*(1), 165-173.

Chan, M. F., Li, J., Bertrand, A., Casbon, A. J., Lin, J. H., Maltseva, I., & Werb, Z. (2013). Protective effects of matrix metalloproteinase-12 following corneal injury. *J Cell Sci*, *126*(17), 3948-3960.

Chiao, Y. A., Zamilpa, R., Lopez, E. F., Dai, Q., Escobar, G. P., Hakala, K., & Lindsey, M. L. (2010). In vivo matrix metalloproteinase-7 substrates identified in the left ventricle post-myocardial infarction using proteomics. *Journal of proteome research*, *9*(5), 2649-2657.

Cohn, J. N., Ferrari, R., & Sharpe, N. (2000). Cardiac remodeling—concepts and clinical implications: a consensus paper from an international forum on cardiac remodeling. *Journal of the American College of Cardiology*, *35*(3), 569-582.

Consili, C., Gatta, L., & Iellamo, F. (2013). Severity of left ventricular dysfunction in heart failure patients affects the degree of serum in duced cardiomyocyte apoptosis. *Importance of inflammatory response and metabolism*, *167*(6), 2859-2866.

DeLeon-Pennell, K. Y., Meschiari, C. A., Jung, M., & Lindsey, M. L. (2017). Matrix metalloproteinases in myocardial infarction and heart failure. In *Progress in molecular biology and translational science* (Vol. 147, pp. 75-100). Academic Press.

Fanjul-Fernández, M., Folgueras, A. R., Cabrera, S., & López-Otín, C. (2010). Matrix metalloproteinases: evolution, gene regulation and functional analysis in mouse models. *Biochimicaet BiophysicaActa (BBA)-Molecular Cell Research*, *1803*(1), 3-19.

Fertin, M., Lemesle, G., Turkieh, A., Beseme, O., Chwastyniak, M., Amouyel, P., ...& Pinet, F. (2013). Serum MMP-8: a novel indicator of left ventricular remodeling and cardiac outcome in patients after acute myocardial infarction. *PloS one*, *8*(8), e71280.

Geurts, N., Martens, E., Van Aelst, I., Proost, P., Opdenakker, G., & Van den Steen, P. E. (2008).β-hematin interaction with the hemopexin domain of gelatinase B/MMP-9 provokes autocatalytic processing of the propeptide, thereby priming activation by MMP-3. *Biochemistry*, *47*(8), 2689-2699.

Gomis-Rüth, F. X. (2009). Catalytic domain architecture of metzincin metalloproteases. *Journal of biological chemistry*, *284*(23), 15353-15357.

Halade, G. V., Jin, Y. F., & Lindsey, M. L. (2013). Matrix metalloproteinase (MMP)-9: a proximal biomarker for cardiac remodeling and a distal biomarker for inflammation. *Pharmacology & therapeutics*, *139*(1), 32-40.

Hu, J., Van den Steen, P. E., Sang, Q. X. A., & Opdenakker, G. (2007). Matrix metalloproteinase inhibitors as therapy for inflammatory and vascular diseases. *Nature reviews Drug discovery*, *6*(6), 480.

Iyer, R. P., de Castro Brás, L. E., Jin, Y. F., & Lindsey, M. L. (2014). Translating Koch's postulates to identify matrix metalloproteinase roles in postmyocardial infarction remodeling: cardiac metalloproteinase actions (CarMA) postulates. *Circulation research*, *114*(5), 860-871.

Johnson, J. L. (2017). Metalloproteinases in atherosclerosis. *European journal of pharmacology*, *816*, 93-106.

Johnson, J. L., 2007. Matrix metalloproteinases: influence on smooth muscle cells and atherosclerotic plaque stability. *Expert Rev. Cardiovasc. Ther.* 5 (2), 265–282.

Kelly, D., Cockerill, G., Ng, L. L., Thompson, M., Khan, S., Samani, N. J., & Squire, I. B. (2007). Plasma matrix metalloproteinase-9 and left ventricular remodelling after acute myocardial infarction in man: a prospective cohort study. *European heart journal*, *28*(6), 711-718.

Kere, U. (2004). Matrix metalloproteinase 28/epilysin expression in cartilage from patients with rheumatoid arthritis and osteoarthritis: comment on the article by Kevorkian et al., *Arthritis & Rheumatism*, *50*(12), 4074-4080.

Kevorkian, L., Young, D. A., Darrah, C., Donell, S. T., Shepstone, L., Porter, S., & Clark, I. M. (2004). Expression profiling of metalloproteinases and their inhibitors in cartilage. *Arthritis & Rheumatism: Official Journal of the American College of Rheumatology*, *50*(1), 131-141.

Kitaoka, H., Kubo, T., Okawa, M., Takenaka, N., Baba, Y., Yamasaki, N., ...& Doi, Y. L. (2011). Plasma metalloproteinase levels and left ventricular remodeling in hypertrophic cardiomyopathy in patients with an identical mutation. *Journal of cardiology*, *58*(3), 261-265.

Klein, T., & Bischoff, R. (2011). Physiology and pathophysiology of matrix metalloproteases. *Amino acids*, *41*(2), 271-290.

Libby, P. (2013). Collagenases and cracks in the plaque. *The Journal of clinical investigation*, *123*(8), 3201-3203.

Lindsey, M. L., & Zamilpa, R. (2012). Temporal and spatial expression of matrix metalloproteinases and tissue inhibitors of metalloproteinases following myocardial infarction. *Cardiovascular therapeutics*, *30*(1), 31-41.

Lindsey, M. L., Iyer, R. P., Zamilpa, R., Yabluchanskiy, A., DeLeon-Pennell, K. Y., Hall, M. E., & Cannon, P. L. (2015). A novel collagen matricryptin reduces left ventricular dilation post-myocardial infarction by promoting scar formation and angiogenesis. *Journal of the American College of Cardiology*, *66*(12), 1364-1374.

Lombardi, R., Betocchi, S., Losi, M. A., Tocchetti, C. G., Aversa, M., Miranda, M., ...& Chiariello, M. (2003). Myocardial collagen turnover in hypertrophic cardiomyopathy. *Circulation*, *108*(12), 1455-1460.

Malgija, B., Rajendran, H. A. D., Maheswari, U., Ebenezer, N. S., Priyakumari, J., &Piramanayagam, S. (2018). Computational analysis of sequential and structural variations in stromelysins as an insight towards matrix metalloproteinase research. *Informatics in Medicine Unlocked*, *11*, 28-35.

Malgija, B., Senthilkumar, N., and Shanmughavel, P. (2018). Collective transcriptomic deregulation of hypertrophic and dilated cardio-myopathy—Importance of fibrotic mechanism in heart failure. *Computational Biology and Chemistry*. 73: 85-94.

Malgija, B., and Shanmughavel, P. (2015). Differential gene expression analysis of hypertrophic and dilated cardiomyopathy signature genes. International *journal of Advanced Research in Computer science and Software Engineering. 5(10)*: 537-543.

Marchenko, N. D., Marchenko, G. N., Weinreb, R. N., Lindsey, J. D., Kyshtoobayeva, A., Crawford, H. C., &Strongin, A. Y. (2004). β-Catenin regulates the gene of MMP-26, a novel matrix metalloproteinase expressed both in carcinomas and normal epithelial cells. *The international journal of biochemistry & cell biology*, *36*(5), 942-956.

McGuire, J. K., Li, Q., & Parks, W. C. (2003). Matrilysin (matrix metalloproteinase-7) mediates E-cadherin ectodomain shedding in injured lung epithelium. *The American journal of pathology*, *162*(6), 1831-1843.

Montero, I., Orbe, J., Varo, N., Beloqui, O., Monreal, J. I., Rodríguez, J. A., ... & Páramo, J. A. (2006). C-reactive protein induces matrix metalloproteinase-1 and-10 in human endothelial cells: implications for clinical and subclinical atherosclerosis. *Journal of the American College of Cardiology*, *47*(7), 1369-1378.

Morine, K. J., Paruchuri, V., Qiao, X., Mohammad, N., Mcgraw, A., Yunis, A., ...& Kapur, N. K. (2016). Circulating multimarker profile of patients with symptomatic heart failure supports enhanced fibrotic degradation and decreased angiogenesis. *Biomarkers*, *21*(1), 91-97.

Murray, M. Y., Birkland, T. P., Howe, J. D., Rowan, A. D., Fidock, M., Parks, W. C., &Gavrilovic, J. (2013). Macrophage migration and invasion is regulated by MMP10 expression. *PloS one*, *8*(5), e63555.

Nagase, H., Visse, R., & Murphy, G. (2006). Structure and function of matrix metalloproteinases and TIMPs. *Cardiovascular research*, *69*(3), 562-573.

Nepomnyashchikh, L. M., Lushnikova, E. L., Bakarev, M. A., Nikityuk, D. B., Yuzhik, E. I., Mzhelskaya, M. M., ... &Karpova, A. A. (2015). Immunohistochemical analysis of MMP-2 expression in the myocardium during the postinfarction period. *Bulletin of experimental biology and medicine*, *159*(4), 505-510.

Nilsson, L., Jonasson, L., Nijm, J., Hamsten, A., & Eriksson, P. (2006). Increased plasma concentration of matrix metalloproteinase-7 in patients with coronary artery disease. *Clinical chemistry*, *52*(8), 1522-1527.

Noji, Y., Shimizu, M., Ino, H., Higashikata, T., Yamaguchi, M., Nohara, A., & Namura, M. (2004). Increased circulating matrix metalloproteinase-2 in patients with hypertrophic cardiomyopathy with systolic dysfunction. *Circulation Journal*, *68*(4), 355-360.

Orbe, J., Rodríguez, J. A., Calvayrac, O., Rodríguez-Calvo, R., Rodriguez, C., Roncal, C., ...& Páramo, J. A. (2009). Matrix metalloproteinase-10 is upregulated by thrombin in endothelial cells and increased in patients with enhanced thrombin generation. *Arteriosclerosis, thrombosis, and vascular biology*, *29*(12), 2109-2116.

Overall, C. M. (2002). Molecular determinants of metalloproteinase substrate specificity: matrix metalloproteinases and new 'intracellular' substrate binding domains, modules and exosites. *MolBiotechnolChem*, *383*, 1059-1066.

Papazafiropoulou, A. and Tentolouris, N., (2009). Matrix metalloproteinases and cardiovascular diseases. *Hippokratia*, *13*(2), 76.

Pérez-Hernández, N., Vargas-Alarcón, G., Martínez-Rodríguez, N., Martínez-Ríos, M. A., Peña-Duque, M. A., de la Peña-Díaz, A., ... & Rodríguez-Pérez, J. M. (2012). The matrix metalloproteinase 2-1575 gene polymorphism is associated with the risk of developing myocardial infarction in Mexican patients. *Journal of atherosclerosis and thrombosis*, 11817.

Petrov, V. V., Fagard, R. H., & Lijnen, P. J. (2002). Stimulation of collagen production by transforming growth factor-β1 during differentiation of cardiac fibroblasts to myofibroblasts. *Hypertension*, *39*(2), 258-263.

Phatharajaree, W., Phrommintikul, A., & Chattipakorn, N. (2007). Matrix metalloproteinases and myocardial infarction. *Canadian Journal of Cardiology*, *23*(9), 727-733.

Pytliak, M., Vaník, V., & Bojčík, P. (2017). Heart Remodelation: Role of MMPs. *The Role of Matrix Metalloproteinase in Human Body Pathologies*, 37.

Rio, M. C. (2005). From a unique cell to metastasis is a long way to go: clues to stromelysin-3 participation. *Biochimie*, *87*(3-4), 299-306.

Rodríguez-Pérez, J. M., Vargas-Alarcón, G., Posadas-Sánchez, R., Zagal-Jiménez, T. X., Ortíz-Alarcón, R., Valente-Acosta, B., ... & Pérez-Hernández, N. (2016). rs3918242 MMP9 gene polymorphism is associated with myocardial infarction in Mexican patients. *Genet Mol Res*, *15*(1), 15017776.

Rutschow, S., Li, J., Schultheiss, H. P., & Pauschinger, M. (2006). Myocardial proteases and matrix remodeling in inflammatory heart disease. *Cardiovascular Research*, *69*(3), 646-656.

Saarialho-Kere, U., KerkelaÈ, E., Suomela, S., Jahkola, T., Keski-Oja, J., & Lohi, J. (2002). Epilysin (MMP-28) expression is associated with cell proliferation during epithelial repair. *Journal of investigative dermatology*, *119*(1), 14-21.

Schönbeck, U., Mach, F., Sukhova, G. K., Atkinson, E., Levesque, E., Herman, M.,. & Libby, P. (1999). Expression of stromelysin-3 in atherosclerotic lesions: regulation via CD40–CD40 ligand signaling in vitro and in vivo. *Journal of Experimental Medicine*, *189*(5), 843-853.

Shi, Y., Zhang, J., Tan, C., Xu, W., Sun, Q., & Li, J. (2015). Matrix metalloproteinase-2 polymorphisms and incident coronary artery disease: a meta-analysis. *Medicine*, *94*(27).

Spinale, F. G. (2002). Matrix metalloproteinases: regulation and dysregulation in the failing heart. *Circulation research*, *90*(5), 520-530.

Spinale, F. G. (2004). *Matrix metalloproteinase gene polymorphisms in heart failure: new pieces to the myocardial matrix puzzle.*

Squire, I. B., Evans, J., Ng, L. L., Loftus, I. M., & Thompson, M. M. (2004). Plasma MMP-9 and MMP-2 following acute myocardial infarction in man: correlation with echocardiographic and neurohumoral parameters of left ventricular dysfunction. *Journal of cardiac failure*, *10*(4), 328-333.

Tamaoki, M., Imanaka-Yoshida, K., Yokoyama, K., Nishioka, T., Inada, H., Hiroe, M., ... & Yoshida, T. (2005). Tenascin-C regulates recruitment of myofibroblasts during tissue repair after myocardial injury. *The American journal of pathology*, *167*(1), 71-80.

Tester, A. M., Cox, J. H., Connor, A. R., Starr, A. E., Dean, R. A., Puente, X. S., ... & Overall, C. M. (2007). LPS responsiveness and neutrophil chemotaxis in vivo require PMN MMP-8 activity. *PloS one*, *2*(3), e312.

Uria, J. A., & Lopez-Otin, C. (2000). Matrilysin-2, a new matrix metalloproteinase expressed in human tumors and showing the minimal domain organization required for secretion, latency, and activity. *Cancer research*, *60*(17), 4745-4751.

Velasco, G., Pendás, A. M., Fueyo, A., Knäuper, V., Murphy, G., & López-Otín, C. (1999). Cloning and characterization of human MMP-23, a new matrix metalloproteinase predominantly expressed in reproductive tissues and lacking conserved domains in other family members. *Journal of Biological Chemistry*, *274*(8), 4570-4576.

Vihinen, P., & Kahari, V. M. (2002). Matrix metalloproteinases in cancer: prognostic markers and therapeutic targets. *International journal of cancer*, *99*(2), 157-166.

Walker, G. A., Masters, K. S., Shah, D. N., Anseth, K. S., & Leinwand, L. A. (2004). Valvular myofibroblast activation by transforming growth factor-β: implications for pathological extracellular matrix remodeling in heart valve disease. *Circulation research*, *95*(3), 253-260.

Wang, J., Xu, D., Wu, X., Zhou, C., Wang, H., Guo, Y., & Cao, K. (2011). Polymorphisms of matrix metalloproteinases in myocardial infarction: a meta-analysis. *Heart*, *97*(19), 1542-1546.

Wang, W. S., Chen, P. M., Wang, H. S., Liang, W. Y., & Su, Y. (2006). Matrix metalloproteinase-7 increases resistance to Fas-mediated apoptosis and is a poor prognostic factor of patients with colorectal carcinoma. *Carcinogenesis*, *27*(5), 1113-1120.

Were, Z., & Gordon, S. A. I. M. O. N. (1975). Elastase secretion by stimulated macrophages - Characterization and regulation. *Journal of Experimental Medicine*, *142*(2), 361-377.

Wilson, E. M., Gunasinghe, H. R., Coker, M. L., Sprunger, P., Lee-Jackson, D., Bozkurt, B., & Spinale, F. G. (2002). Plasma matrix metalloproteinase and inhibitor profiles in patients with heart failure. *Journal of cardiac failure*, *8*(6), 390-398.

Wilson, E. M., Moainie, S. L., Baskin, J. M., Lowry, A. S., Deschamps, A. M., Mukherjee, R., ... & Gorman, R. C. (2003). Region-and type-specific induction of matrix metalloproteinases in post–myocardial infarction remodeling. *Circulation*, *107*(22), 2857-2863.

Yabluchanskiy, A., Li, Y., J Chilton, R., & L Lindsey, M. (2013a). Matrix metalloproteinases: drug targets for myocardial infarction. *Current drug targets*, *14*(3), 276-286.

Yabluchanskiy, A., Ma, Y., Iyer, R. P., Hall, M. E., & Lindsey, M. L. (2013b). Matrix metalloproteinase-9: many shades of function in cardiovascular disease. *Physiology*, *28*(6), 391-403.

Zhao, Y. G., Xiao, A. Z., Park, H. I., Newcomer, R. G., Yan, M., Man, Y. G., ... & Sang, Q. X. A. (2004). Endometase/matrilysin-2 in human breast ductal carcinoma in situ and its inhibition by tissue inhibitors of metalloproteinases-2 and-4: a putative role in the initiation of breast cancer invasion. *Cancer research*, *64*(2), 590-598.

Zile, M. R., DeSantis, S. M., Baicu, C. F., Stroud, R. E., Thompson, S. B., McClure, C. D., ... & Spinale, F. G. (2011). Plasma biomarkers that reflect determinants of matrix composition identify the presence of left ventricular hypertrophy and diastolic heart failure. *Circulation: Heart Failure*, *4*(3), 246-256.

BIOGRAPHICAL SKETCH

M. Beutline Malgija

Affiliation: Research Associate

Education: MSc, PhD

Research and Professional Experience:

- Research associate at Bioinformatics Centre of BTISnet, Madras Christian College
- Worked as Teaching assistant in Kerala Agricultural University
- Awarded doctorate for the research work entitled "Gene expression and Gene regulatory network data integration- An insight into molecular mechanism of Cardiovascular Disease." From Bharathiar University, Coimbatore, India.

Honors: Received Gold medal in Masters.

Publications from the Last 3 Years:

Beutline Malgija M and Shanmughavel P. A review on cardiovascular genetics and its in silico methods. *Indian Journal of Biotechnology*. 2018, 17: 205-210.

Beutline Malgija M, Host Antony David Rajendran, Sylvester Darvin, Natchimuthu Santhi and Joyce Priyakumari. In Silico Exploration of HIV Entry Co-receptor Antagonists: A combination of Molecular Modeling, Docking and Molecular Dynamics Simulations. *Acta Scientific Pharmaceutical Sciences*. 2019, 3(2): 60-67.

Beutline Malgija M, Host Antony David, Uma maheswari M, Nivetha Sarah Ebenezer, Shanmughavel P and Joyce Priyakumari. "Computational analysis of Sequential and Structural variations in Stromelysins as an Insight towards Matrix Metalloproteinase research." *Informatics in Medicine Unlocked*. 2018, 11: 28-35.

Beutline Malgija M, Senthilkumar N and Shanmughavel P. Collective transcriptomic deregulation of hypertrophic and dilated cardiomyopathy—Importance of fibrotic mechanism in heart failure. *Computational Biology and Chemistry*.2018, 73: 85-94.

Host Antony David Rajendran, Beutline Malgija M, Nivetha Sarah Ebenezer, Uma Maheswari, Victor Roch G, Joyce Priyakumari and Savarimuthu Ignacimuthu. Homology modeling and molecular docking studies of Purple acid Phosphatase from *Setaria italica* (Foxtail millet). *International Journal of Scientific Research in Biological Sciences*. 2018, 5(4): 119-124.

In: A Closer Look at Metalloproteinases ISBN: 978-1-53616-517-3
Editor: Lena Goodwin

Chapter 4

THE ROLE OF MATRIX METALLOPROTEINASES IN THE PATHOGENESIS OF PSORIASIS

Julia A. Mogulevtseva*[1], *Alexandre Mezentsev*[2,*], *PhD* and *Sergey A. Bruskin*[2], *PhD
[1]Department of Agronomy and Biotechnology,
K. A. Timiryazev Russian State Agricultural University (RSAU-MAA),
Russian Ministry of Science and Higher Education, Moscow, Russia
[2]Department of Functional Genomics,
N. I. Vavilov Institute of General Genetics,
Russian Academy of Sciences, Moscow, Russia

ABSTRACT

The aim of this paper was to summarize the data about the role of matrix metalloproteinases in the pathogenesis of psoriasis. In psoriasis, matrix metalloproteinases facilitate structural remodeling of the epidermis that, in turn, results in the development of psoriatic plaques. Particularly, they influence the composition of the extracellular matrix and regulate the strength of intracellular contacts between the skin cells. In the dermis, the

[*] Corresponding Author's E-mail: mesentsev@vigg.ru.

interaction of matrix metalloproteinases with vascular endothelial cells results in reshaping and vasodilation of venous capillaries. After all, microcapillaries become more permeable for immune cells. At the molecular level, the expression of *MMP1*, *MMP9* and *MMP12* correlates with disease severity. It increases with disease progression and decreases in remission. The separate sections of the paper are dedicated to the role of matrix metalloproteinases in generation of matrikines from the proteins of extracellular matrix and regulation of cytokines that are involved in the pathogenesis of psoriasis. Thus, we assessed the contribution of matrix metalloproteinases to the pathogenesis of psoriasis and analyzed changes in physiological processes caused by their differential expression in lesional skin.

Keywords: matrix metalloproteinases, psoriasis, inflammation, matrikines, angiogenesis

ABBREVIATIONS

BM	basal membrane;
DS	desmosomes;
IL	interleukin;
ING	interferon γ;
ECM	extracellular matrix;
MMP	matrix metalloproteinase;
TGFβ	transforming growth factor β;
TNF	tumor necrosis factor α;
CTGF	connective tissue growth factor;
VEGF	vascular endothelial growth factor;
HD	hemidesmosomes;
HB-GAM	heparin-binding brain mitogen/pleiotrophin;
MCP	monocyte chemoattractant protein

INTRODUCTION

Matrix metalloproteinases (MMPs) are a group of zinc-containing, calcium-dependent endopeptidases (Cerofolini, Fragai, and Luchinat 2018). Similarly to the other endopeptidases, MMPs cleave peptide bonds in the molecules of other proteins (substrate proteins). Although a typical active center of the enzyme contains two zinc atoms, one of them is needed to maintain MMP in a catalytically active state. In turn, the second zinc atom is directly involved in catalysis. The catalytic zinc interacts with the molecule of water, which is also located in the active center. Due to their interaction, water transforms into hydroxyl anion: OH, which is a strong nucleophilic group. Then, the hydroxyl anion attacks the nearest peptide bond of the substrate (Figure 1) and cleaves the substrate protein into two peptides of lower molecular weights.

Since MMPs have different substrate specificities, it would be convenient to split these proteins into several subgroups of smaller sizes, such as collagenases (MMP1, -8, and -13), gelatinases (MMP2 and -9), stromelysins (MMP3, -10 and -11), matrilizines (MMP7 and -26), membrane-bound metalloproteinases (MMP14, -15, -16, -17, -23A, -23B, -24 and -25), and other metalloproteinases (MMP12, -19, -20, -21, -27 and -28). In this respect, 23 different MMPs are encoded in the human genome and one of the genes, the gene encoding MMP23, is represented in the genome by two identical copies – MMP23A and -23B.

MMPs are irreplaceable for normal physiological processes. In the extracellular matrix (ECM), MMPs cleave most of its protein components. They also activate cytokines, cellular receptors and proenzymes. MMPs contribute to cell division and migration, the interaction of the cells (adhesion) and their differentiation (Cui, Hu, and Khalil 2017). For this reason, it is not surprising that many MMPs are directly involved in the inflammatory response because it would not progress if any of the named processes and functions were disabled.

Psoriasis is one of the most abundant dermatoses (Lebwohl 2018). According to the National Foundation of Psoriasis (USA), ~ 125 million people, i.e., 2-3% of the total population is diagnosed with psoriasis.

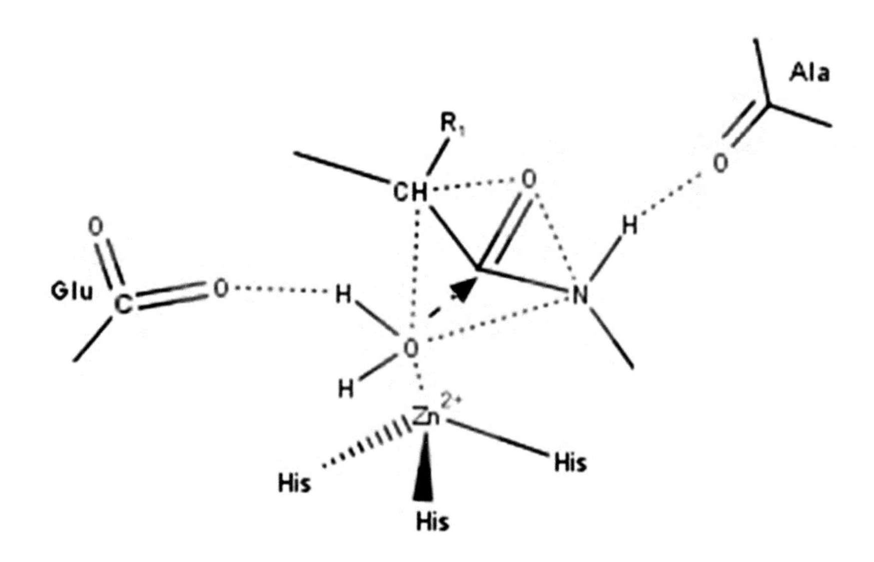

Figure 1. Schematic representation of the reaction catalyzed by matrix metalloproteinases. Interaction between the oxygen atom of water and the catalytic zinc atom of MMP results in formation of hydroxyl anion (:OH-). Then, the hydroxyl anion (:OH-) attacks the nearest α-carbon atom of substrate catalyzing the hydrolysis of the peptide bond (-C-N-).

In Russia, the incidence of psoriasis is ~ 1.9%, and the total number of psoriasis patients is 2.8 million (Khamaganova et al. 2015). The hallmark of psoriasis is scaling red patches often referred to as plaques. The plaques are developed due to epidermal hyperplasia and chronic inflammation in the diseased skin. In 23-27% and 30% of cases, respectively, the disease also targets nails and joints (Ritchlin, Colbert, and Gladman 2017, Dopytalska et al. 2018). Moreover, the psoriasis patients are more likely to be diagnosed with comorbidities, such as cardiovascular disease, type II diabetes, atherosclerosis, etc. (Takeshita et al. 2017).

The appearance of psoriatic plaques on the skin is accompanied by penetration of the skin with immune cells. Immune cells secrete pro-inflammatory cytokines, such as tumor necrosis factor α (TNF), interleukin 17 (IL-17) and interferon γ (ING) to ECM. Their secretion leads to the hyperproliferation of epidermal keratinocytes, it shortens their life cycle and modifies their differentiation program. The named changes cause a structural remodeling of the epidermis. In the first turn, MMPs modify the intercellular contacts and change the composition of ECM (Mezentsev, Nikolaev, and Bruskin 2014). They also facilitate the penetration of dermal microcapillaries by immune cells and regulate the biological activity of some cytokines. For example, MMP1, -2, -3, -7, -9, -12, -13 and -17 proteolytically activate the latent form of TNF. Then, TNF disconnects from the cellular membrane, solubilizes and becomes biologically active (Nissinen and Kahari 2014). For these reasons, it would be important for us to know the substrate specificity of the individual MMPs and be able to control their expression in lesional psoriatic skin. In this respect, the aim of this study was to evaluate the role of MMPs in the pathogenesis of psoriasis.

Expression of MMPs in the Skin of Healthy Individuals and Psoriasis Patients

The analysis of gene expression by microarray reveals an induction of *MMP1*, *-9* and *-12* in lesional psoriatic skin (lesional skin) (Mitsui et al. 2012). Their expression correlates with severity of the disease. In contrast, the therapy gradually normalizes their expression, unless the patents achieve remission of the disease (Buommino et al. 2012, Cordiali-Fei et al. 2006, Starodubtseva et al. 2011). In turn, microdissection of the skin with laser let separate lesional epidermis from dermis and reveal that the induction *MMP1*, *-9* and *-12* occurs in the epidermis whereas in the dermis, their expression remains at the same level (Mitsui et al. 2012). In addition, the authors report that, in the epidermis, the expression of two other *MMPs*, *MMP27* and *-28*, significantly decreases.

The results of microarray analysis are in line with the data obtained by the methods of immunohistochemistry and immunoblotting at the post-translational level. Firstly, these data suggest that many MMPs cannot be detected in either the skin of healthy individuals (healthy skin), or healthy-looking skin of psoriasis patients (unaffected skin). In this respect, the list of MMPs expressed in healthy skin is limited to MMP2 (Fleischmajer et al. 2000, Simonetti et al. 2006), MMP3 (Saarialho-Kere et al. 1994), MMP14

(Fleischmajer et al. 2000, Hitchon et al. 2002), MMP15 (Hitchon et al. 2002, Uhlen et al. 2015), MMP16 (Uhlen et al. 2015), MMP19 (Sadowski et al. 2003, Impola et al. 2003, Suomela et al. 2003) and MMP28 (Saarialho-Kere et al. 2002, Reno et al. 2005).

Particularly, MMP2 is located in the basal layer of the epidermis (Simonetti et al. 2006) whereas transmembrane MMPs (MMP15 and -16) are detected in basal and suprabasal layers (Uhlen et al. 2015) (Figure 2). MMP16 is expressed by keratinocytes and melanocytes. In contrast, its expression is usually at low level in dermal fibroblasts (Uhlen et al. 2015). In turn, the expression of MMP3 (Saarialho-Kere et al. 1994), MMP10 (Kerkela et al. 2001), MMP19 (Sadowski et al. 2003, Impola et al. 2003, Suomela et al. 2003) and MMP28 (Saarialho-Kere et al. 2002, Reno et al. 2005) intensifies following a skin damage. In wounded skin, these MMPs are detected in basal and suprabasal layers of the epidermis. In the dermis, MMP3 is expressed by fibroblasts (Brinckerhoff et al. 1992, Lindqvist, Phil-Lundin, and Engstrom-Laurent 2012). In contrast, fibroblasts are MMP10 (Madlener et al. 1996) and MMP19 (Suomela et al. 2003) negative if stained with antibodies directed to these proteins. Unlike MMP3, MMP19 is expressed in vascular endothelium of the primarily lymphatic vessels and microcapillaries supplying the blood to the glands and hair follicles (Suomela et al. 2003). In the sebaceous glands, MMP19 is synthesized in undifferentiated cells of the secretary epithelium (Sadowski et al. 2003, Suomela et al. 2003). In the sweat glands, the staining profile for MMP19 is diffusive. In this respect, it would be difficult to identify the cells that expressed MMP19. In the hair follicles, MMP19 is found in the outer root vagina. Finally, there is no evidence of MMP28 expression in the dermis (Kuivanen et al. 2005).

In psoriatic plaques, the distribution of MMPs significantly changes (Figure 2). Particularly, MMP1 (Flisiak, Porebski, and Chodynicka 2006, Wolk et al. 2014), MMP3 (Lindqvist, Phil-Lundin, and Engstrom-Laurent 2012), MMP9 (Simonetti et al. 2006, Feliciani et al. 1997) and MMP12 (Suomela et al. 2001) are induced and their staining profiles with specific antibodies significantly overlap to each other. In the epidermis, they are detected in the basal layer. Moreover, MMP9 expression correlates with the intensity of the inflammatory response. In some cases, MMP9 distribution in lesional skin may look diffusive, i.e., MMP9 presence is not limited to the basal layer and MMP9 positive cells can be observed in the suprabasal layer (Feliciani et al. 1997). Moreover, a small number of epidermal keratinocytes express MMP26. These MMP26 positive cells are specifically associated with the basal layer. They can be located in the upper part of rete ridges (Ahokas et al. 2005). Besides MMP9, the suprabasal keratinocytes also express MMP2 (Fleischmajer et al. 2000, Simonetti et al. 2006) and MMP19 (Sadowski et al. 2003, Suomela et al. 2003). For instance, a small number of cells expressing MMP19 can be found in the upper part of rete ridges.

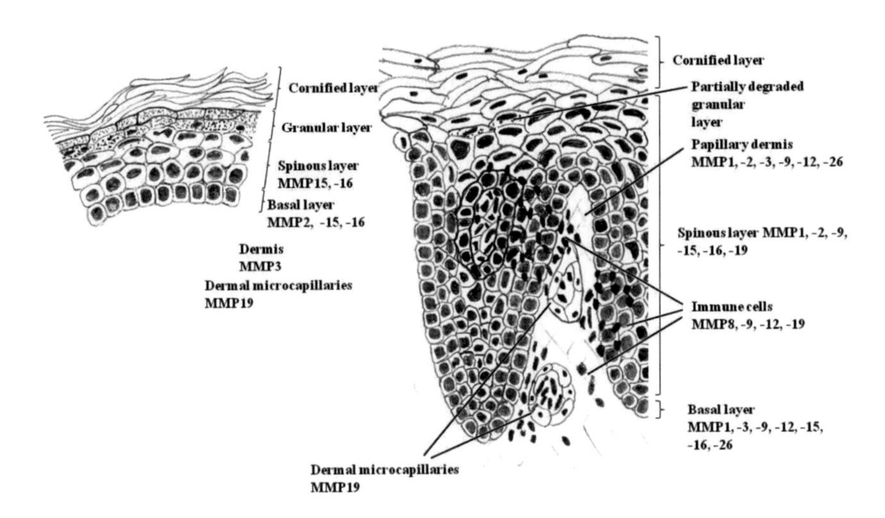

Figure 2. Localization of matrix metalloproteinases in healthy and lesional skin. Healthy and lesional skin are shown in left and right plots, respectively.

In the dermis, MMP1 (Wolk et al. 2014), MMP2 (Feliciani et al. 1997), MMP3, -9 (Lindqvist, Phil-Lundin, and Engstrom-Laurent 2012), and MMP12 (Suomela et al. 2001) can be detected in the dermal papillaries; as well as in the areas surrounding the blood vessels, whereas the location of MMP19 does not change significantly compared to the dermises of healthy and unaffected skin (Sadowski et al. 2003, Suomela et al. 2003). Furthermore, the antibodies specific to the certain MMPs react with immune cells infiltrated lesional skin. For instanse, macrophages are positive for

MMP9, -12 and -19 (Suomela et al. 2003), whereas neutrophils interact with antibodies directed to MMP8 (Tsaousi et al. 2016) and MMP9 (Suomela et al. 2001). In addition to macrophages, MMP19 is secreted by the helper cells, T_{h1}, that play a crucial role in the pathogenesis of psoriasis (Sedlacek et al. 1998).

In conclusion, similarly to microarray analysis, the results of immunohistochemistry experiments reveal disease-associated changes in the expression of individual MMPs in lesional and unaffected skin. Particularly, MMP2 and -15 expression levels in healthy skin are lower than ones in unaffected skin and their expression levels in unaffected skin are lower than ones in lesional skin (Fleischmajer et al. 2000, Hitchon et al. 2002). In contrast, changes in MMP14 expression occur in the opposite direction (Hitchon et al. 2002). On the other hand, the immune-histochemistry is not sensitive enough to detect MMP1, -9 and -12 in healthy and unaffected skin (Simonetti et al. 2006, Flisiak, Porebski, and Chodynicka 2006, Wolk et al. 2014, Feliciani et al. 1997, Suomela et al. 2001).

The Role of MMPs in Stabilizing the Inflammatory Response and Their Interaction with Chemokines and Cytokines

Chronic skin inflammation is one of the most prominent features of psoriasis. Migration of immune cells to lesional skin leads to their accumulation in the plaques. In total, lesional skin may accumulate up to 20 of 28-30 million leukocytes that are normally present in human body (Saalbach et al. 2008). A typical psoriatic infiltrate contains at least four different types of cells – lymphocytes, macrophages, dendritic cells and neutrophils. Importantly, the cellular composition of the infiltrate at different stages of the disease is not the same (Christophers and Mrowietz 1995). Particularly, the ratio of macrophages reaches the maximum at the time when the disease exacerbates and coincides with an active growth of psoriatic plaques.

The ratio of $CD8^+$ Tc cells decreases whereas the ratio of $CD4^+$ T cells, such as T_{h1}, T_{h17} and T_{h22}, increases when the growth of the plaques slows down. In turn, the ratio of neutrophils increases in the so-called. "late" infiltrate after the inflammatory process is already stabilized.

Importantly, the mentioned cells are preferably located in different parts of lesional skin because they secrete different proteolytic enzymes, including MMPs (Bar-Or et al. 2003, Chou, Chan, and Werb 2016). Particularly, $CD4^+$ T_h and dendritic cells are more likely to be found in the epidermis, whereas $CD8^+$ T cells and neutrophils - in the dermis (Brandner et al. 2015). These differences help the immune cells to adapt to their new environment and, in turn, regulate the intensity of the immune response. Moreover, some proinflammatory cytokines, are susceptible to MMPs (Butler and Overall 2009). Processing cytokines, MMPs contribute to the regulation of the inflammatory response and influence the cellular composition of the immune infiltrate in lesional skin.

MMPs can act on cytokines either directly or indirectly. In the first case, MMPs cleave them similarly to the other substrates. In turn, the truncated cytokines may either acquire or lose their biological activity. In the second case, MMPs cleave an intermediate molecule, which is not a cytokine. However, its truncated form can cause changes in biological activities and bioavailabilities of certain cytokines.

For instance, MMPs directly inactivate monocyte chemotactic proteins (MCP1, -2, 3 and -4) (McQuibban et al. 2002, McQuibban et al. 2000). In this case, the proteolytic enzyme cleaves a small peptide from the N-terminus of the molecule. Unlike the non-truncated protein, which binds to the specific receptor, activates intracellular signaling mechanisms and stimulates the cell to migrate, the truncated protein, which still preserves the ability to interact with the receptor, looses the abilities to stimulate cell signaling and influence the cell behavior.

In contrast, some other proteolytically digested cytokines gain their biological activity. For instance, the proteolysis of transforming growth factor β (TGFβ), TNF or CXCL8 results in their activation (Table 1).

Table 1. Role of human matrix metalloproteinases in the proteolysis of chemokines and cytokines

Substrate	Enzymes	Effect	References	Substrate	Enzymes	Effect	References
CCL1	MMP3, -12,-14	n.d.	(Starr et al. 2012)	CXCL3	MMP12	↓	(Dean et al. 2008)
CCL2/MCP1	MMP1, -3, -8, -9, -12	↓	(Starret al. 2012)	CXCL5/ENA78	MMP8, -9, -12	↓ - MMP9 and -12 ↑ - MMP8	(Dean et al. 2008, Van Den Steen et al. 2003)
CCL3	MMP1, -8, -9	↓	(Starr et al. 2012)	CXCL8/IL8	MMP8, -9, -12	↓ - MMP12 ↑ - MMP8 and -9	(Dean et al. 2008, Van Den Steen et al. 2003,Li et al. 2004)
CCL4	MMP1, -2, -3, -8, -9, -12, -14	n.d.	(Starret al. 2012)	CXCL12/SDF1	MMP1, -2, -3, -9, -14	↓	(McQuibban et al. 2001)
CCL7/MCP3	MMP1, -2, -3, -12, -14	↓	(Starr et al. 2012)	CXCL6/GCP2	MMP8, -9	≈	(Van Den Steenet al. 2003)
CCL8/MCP2	MMP1, -2, -3, -8, -12,-14	↓	(Starret al. 2012)	HB-EGF	MMP3	↑	(Suzuki et al. 1997)
CCL11	MMP2, -12	↓	(Starr et al. 2012)	IL-1β	MMP1, -2, -3, -9	↓	(Ito et al. 1996)
CCL13/MCP4	MMP1, -2, -3, -8, -9, -12, -14	↓	(Starr et al. 2012)	TGFβ	MMP2, -3, -9	↑	(Yu and Stamenkovic 1995, Maeda et al. 2002)
CX3CL1	MMP2, -12	↓ - MMP2 ↑ - MMP12	(Dean et al. 2008, Bourd-Boittin et al. 2009)	TNF	MMP1, -2, -3, -9	↑	(Gearing et al. 1994)
CXCL2	MMP12	↓	(Dean et al. 2008)	VEGF	MMP1, -3, -9, -19	↑	(Lee et al. 2005)

Symbols ↑ and ↓ indicate that the biological activity of generated peptides is higher and lower, respectively, than one of the substrate, n.d. – not defined. Symbol ≈ indicates that the biological activities of the generated peptide and the substrate are about the same.

In this case, MMPs interact with a cytokine bound to the cell membrane and cleave it off. After proteolysis, the truncated protein becomes water-soluble, it disperses in ECM, enters the bloodstream and creates a concentration gradient for immune cells.

In turn, MMP2 and -9-catalyzed degradation of inhibitory complexes of vascular endothelial growth factor (VEGF) with pleiotrophin (HB-GAM) and connective tissue growth factor (CTGF) is an example of how MMPs can activate cytokines and chemokines indirectly (Dean et al. 2007). Earlier, it was shown that ECM of healthy skin serves as a depository for some growth factors (Schonherr and Hausser 2000). Particularly, the secreted VEGF is mainly stored in inhibitory complexes with HB-GAM and CTGF. The beginning of structural rearrangements in the epidermis and the development of the plaques coincide with induction of certain MMPs. In turn, MMPs, such as MMP2 and MMP9, proteolytically inactivate CTGF and HB-GAM, releasing the biologically active VEGF from the inhibitory complexes. The accumulation of biologically active VEGF in ECM and its following interaction with blood vessel cells lead to neovascularization of the inflammatory skin (Bergers et al. 2000).

The Role of MMPs in the Degradation of Intercellular Contacts and the Biosynthesis of Matrikines

A release of secretory granules into the lumen of dermal microcapillaries is one of the early events in the development of psoriatic plaques (Chiricozzi et al. 2018). The chemokines released from these granules stimulate the migration of macrophages to the dermis. After penetrating the blood vessel wall, macrophages migrate to the papillary dermis. In turn, lymphocytes and neutrophils appear in the dermis following the macrophages. Secreting the proteolytic enzymes, immune cells reach the basal membrane (BM) that separates the dermis from the epidermis.

MMPs that exert a broader substrate specificity compared to the other proteases are able to cleave as BM proteins as well as the proteins that contribute to the formation of intercellular contacts in the epidermis, primarily hemihedemosomes and desmosomes, (Chermnykh, Kalabusheva, and Vorotelyak 2018) (Table 2). Respectively, different immune cells exhibit different MMPs expression patterns. T-lymphocytes predominantly express MMP2, -7, -8, -9, -10, -11, -14, -15, -16, -19, -21, -23, -24, -26 and -28 (Bar-Or et al. 2003), whereas macrophages and neutrophils express MMP1, -2, -3, -7, -8, -9, -10, -12, -13, -14, -16, -19, -25, -26 and MMP1, -8, -9, -10, -25, -26, respectively (Chou, Chan, and Werb 2016).

A degradation of desmosomes makes possible for immune cells enter the suprabasal epidermis. Due to a partial degradation of ECM proteins (Table 2), the intercellular space increases facilitating a further migration of immune cells between the layers of partially differentiated epidermal keratinocytes. Despite of hypogranulosis, i.e., a partial degradation of the granular layer, epidermal keratinocytes of lesional epidermis remain interconnected by tight junctions that are not susceptible to MMPs. Respectively, lesional epidermis preserves its barrier function, similarly to the epidermis of healthy skin (Brandner et al. 2015). This phenomenon can be explained at least by two reasons. First, the protein compositions of tight junctions in lesional and healthy skin are not the same. Compared to tight junctions of healthy skin, tight junctions of lesional skin contain less occludin, claudin 1 (Kirschner et al. 2009, Watson et al. 2007) and 7 (Kirschner et al. 2009). In turn, ZO-1 protein and claudin 4 are differentially distributed between the epidermal layers. Second, the expression of tight junctions proteins and the formation of tight junctions in lesional skin starts earlier, compared to healthy skin. In healthy skin, tight junctions proteins are detected in the granular layer of the epidermis, whereas, in lesional skin, they are already expressed in the upper part of the spinous epidermis (Kirschner et al. 2009).

Table 2. Role of human matrix metalloproteinases in the proteolysis of extracellular matrix proteins

Substrate	Enzymes	Location
aggrecan	MMP1, -2, -3, -7 and -9 (Fosang et al. 1993), MMP8 (Fosang et al. 1994), MMP12 (Durigova et al. 2011), MMP13 (Fosang et al. 1996), MMP14 and -15 (d'Ortho et al. 1997), MMP19 (Stracke et al. 2000)	ECM
versican	MMP7 (Halpert et al. 1996)	ECM
vitronectin	MMP1, -2, -3, -7 and -9 (Imai et al. 1995), MMP12 (Sternlicht and Werb Z 2001), MMP14 (Ohuchi et al. 1997), MMP26 (Marchenko et al. 2001)	ECM
desmoglein 2	MMP9 (Butin-Israeli et al. 2016)	DS
desmoglein 3	MMP2 and -9 (Cirillo et al. 2007)	DS
E-cadherin	MMP2 (Li et al. 2010), MMP3 and -7 (Noe et al. 2001)	DS
β-catenin	MMP7 (Rims and McGuire 2014)	
collagen I	MMP1 (Pilcher et al. 1997), MMP2 (Aimes and Quigley 1995), MMP8 (Van Wart 1992), MMP9 (Okada et al. 1995), MMP14 (Gioia et al. 2007), MMP19 (Sternlicht and Werb 2001)	ECM
collagen IV	MMP2 (Eble et al. 1996), MMP9 (Chiu and Lai 2014)	HD
collagen VII	MMP1 and -2 (Seltzer et al. 1989), MMP9 (Birkedal-Hansen 1995)	ECM
collagen XVII	MMP9 (Stahle-Backdahl 1994)	HD
collagen XVIII	MMP3, -9, -12 and -13 (Ferreras et al. 2000), MMP7 (Lin et al. 2001, Chang et al. 2005), MMP14 (Chang et al. 2005)	HD
laminin 5	MMP2 and -14 (Oku et al. 2006), MMP7 (Yamamoto et al. 2010), MMP12 (Mydel et al. 2008), MMP13 (Zhao et al. 2015)	HD
laminin 10	MMP14 (Bair et al. 2005)	HD
nidogens	MMP3 and -7 (Mayer et al. 1993), MMP12 (Gronski et al. 1997), MMP14 and -15 (d'Ortho et al. 1997), MMP19 (Stracke et al. 2000)	ECM
perlecan	MMP2 and -9 (Pearce and Shively 2006), MMP14 and -15 (d'Ortho et al. 1997)	ECM
syndecan 4	MMP13 (Zhang et al. 2010)	ECM
tenascin C	MMP2, -3 and -7 (Siri et al. 1995), MMP13 (Knauper et al. 1997), MMP14 and -15 (d'Ortho et al. 1997), MMP19 (Stracke et al. 2000)	ECM
fibrillin I	MMP2, -3, -9, -12 and -14 (Ashworth et al. 1999)	ECM

Table 2. (Continued)

Substrate	Enzymes	Location
fibrillin II	MMP3 (Sasaki et al. 1996)	ECM
fibronectin	MMP1 and -3 (Zhang et al. 2012), MMP2 (Jiao et al. 2012), MMP7 (Yamamoto et al. 2010), MMP12 (Gronski et al. 1997), MMP13 (Knauper et al. 1997), MMP14 (Ohuchi et al. 1997), MMP15 (d'Ortho et al. 1997), MMP19 (Stracke et al. 2000), MMP25 (Kang et al. 2001), MMP26 (Marchenko et al. 2001)	ECM
elastin	MMP2 (Senior et al. 1991), MMP7, -9 and -12 (Heinz et al. 2010)	ECM
CD44	MMP14 (Kajita et al. 2001)	ECM
CD138	MMP7 (Li et al. 2002), MMP14 (Endo et al. 2003)	ECM

DS – desmosomes; ECM – extracellular matrix; HD – hemidesmosomes.

Degrading ECM proteins, MMPs generate biologically active peptides known as matrikines. The primary role of matrikines is to regulate migration and proliferation of immune cells (Ricard-Blum and Vallet 2017). Most of the MMP-generated matrikines originate from a relatively small group of cutaneous ECM proteins. This group consists elastin, tenascin, collagens I, IV and XVIII and laminins. Proteolytic degradation of elastin by MMP7, MMP9, and MMP12 (Heinz et al. 2010) leads to the generation of several dozen elastin fragments, some of which, like VGVAPG, serve as chemoattractants to macrophages (Houghton et al. 2006) and monocytes (Senior et al. 1984). Furthermore, some elastin fragments increase the proliferation rate of vascular endothelial cells (Mochizuki, Brassart, and Hinek 2002), stimulate the differentiation of naive T cells into T_{h1} lymphocytes (Debret et al. 2005) and induce the genes encoding *MMP2* and *-14* (Ntayi et al. 2004). In turn, collagen IV is a major source of proline-containing oligopeptides (e.g., Pro-Gly-Pro), which are chemoattractants of neutrophils (Weathington et al. 2006). Moreover, they slow down angiogenesis (Crunkhorn 2017).

In turn, the biologically active peptides originated from laminin 5 in the reaction catalyzed by either MMP2 or -14 serve as chemotactics to neutrophils recruiting them to the site of inflammation (Mydel et al. 2008). Endostatin and endostatin-like peptides produced by MMP2, -3, -7, -12, -13, or -14 from collagen XVIII (Lin et al. 2001, Ferreras et al. 2000) as well as endorepelin produced by either MMP2 or -9 from perlecan (Pearce and Shively 2006) exhibit a potent angiostatic effect (Ramakrishnan et al. 2007, Gubbiotti, Neill, and Iozzo 2017). In addition, we would like to mention the tripeptides *Arg-Gly-Asp* (RGD) and *Gly-His-Lys* (GHK), which are originated from type I collagen by either MMP1 or MMP13. The first inhibits cell adhesion (Korff and Augustin 1999) and suppresses angiogenesis initiating apoptosis in vascular endothelial cells (Meerovitch et al. 2003). The copper complex of the second, *Gly-His-Lys*-Cu, serves as a chemotactic to monocytes, macrophages and mast cells (Poole and Zetter 1983, Zetter, Rasmussen, and Brown 1985). This compound also stimulates angiogenesis and deposition of ECM in vivo (Pickart and Margolina 2018).

In conclusion, we would like to acknowledge that discovery and characterization of new matrikines is an ongoing process. To date, many matrikines remain unidentified, and some physiological effects of known matrikines are not fully characterized. In the same time, matrikines are a class of biologically active compounds that has a huge therapeutic potential. In the future, matrikines could be used to inhibit angiogenesis and suppress the recruitment and homing of immune cells to the inflamed tissue. The later could be beneficial for the patients diagnosed with autoimmune diseases, such as psoriasis.

THE ROLE OF MMPS IN NEOVASCULARIZATION

Functioning of the blood vessels to a certain extent depends on the balance between catalytically active MMPs and their inhibitors. In lesional skin, this balance is shifted due to a higher proteolytic activity promoting angiogenesis as well as remodeling of the existing blood vessels. Moreover, dermal capillaries are dilated in lesional skin compared to ones in healthy dermis (Stinco, Lautieri, and Patrone 2006).

In healthy skin, arterial and venous capillaries are symmetrical and their length is about the same (Figure 3). In turn, the development of psoriatic plaques is accompanied by elongation of venous capillaries. The modified part of the capillary loop remains in parallel to BM, which separates the epidermis from the dermis (Micali et al. 2010). The total capillary surface increases by factor 4 (Creamer et al. 1997, Hern et al. 1999) and the diameter increases by factor 2.5 (Stinco, Lautieri, and Patrone 2006). In this respect, the blood flow in the remodeled capillaries intensifies by factor 11, compared to one in healthy skin or 3.5 compared to unaffected skin of the patients (Hern et al. 1999). In addition, the capillary wall is perforated by 0.03 to 1.9 μm slits. These slits facilitate the penetration the capillary walls by immune cells.

Remodeling the dermal microcapillaries follows the changes in proliferation and migration of endothelial cells (Braverman 2000). The role of MMPs in this process is the following.

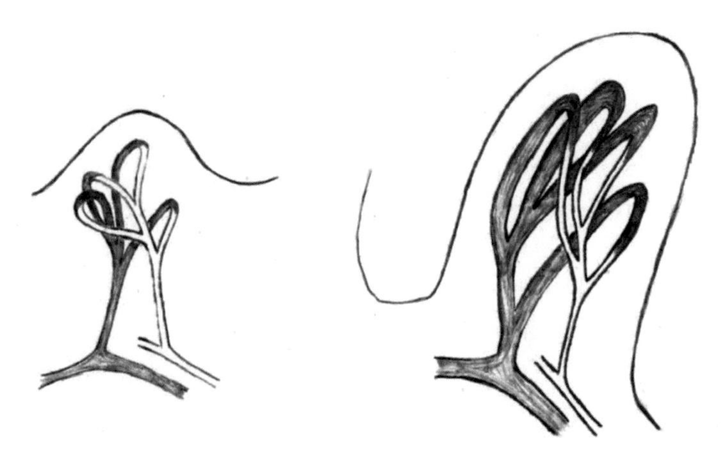

Figure 3. Schematic representation of changes in dermal capillaries caused by psoriasis. Healthy and diseased capillaries are shown in left and right plots, respectively. Venous ends of the capillaries are colored in grey and arterial ends are left uncolored as depicted in the figure.

First, they cleave the proteins of ECM that surround the capillaries. Second, MMPs facilitate the migration of microvessel endothelial cells. Third, MMPs activate cytokines (e.g., TNF), those latent forms are associated with the cell membrane. Fourth, MMPs secreted by the cells of the vascular endothelium, such as MMP1, -2, and -9, as well as transmembrane metalloproteinases, play a decisive role in the degradation of the BM and the connective tissue that surrounds the capillary (Siefert and Sarkar 2012). Fifth, MMPs secreted by activated immune cells facilitate their penetration of BM and capillary walls. This process requires a secretion of MMP1 and -9 that cleave collagen, laminin and elastin.

Conclusion

Despite MMPs are one of the most well studied groups of proteolytic enzymes and they play a crucial role in the pathogenesis of many autoimmune disorders, only few MMPs inhibitors are approved for clinical use (Mezentsev, Nikolaev, and Bruskin 2014). Presumably, polyfunctionality of MMPs and their involvement in many physiological processes make them a difficult therapeutic target and, in this respect, the design of their specific inhibitors becomes a challenging task. On the other hand, the effectiveness of MMPs inhibitors can be improved if the candidate drugs would act locally and rapidly degrade in bloodstream. In this paper, we showed that MMPs play an important role in the formation of psoriatic plaques, initiation and maintenance of the inflammatory response, as well as structural rearrangement of the dermal microcapillaries. As we believe, new "local" MMPs inhibitors with minimized systemic effect could be used to abolish the inflammatory response and be especially beneficial for many patients who periodically experience an exacerbation of their psoriasis.

REFERENCES

Ahokas, K. T., Skoog, S., Suomela, L., Jeskanen, U., Impola, K., Isaka and Saarialho-Kere, U. (2005). Matrilysin-2 (matrix metalloproteinase-26) is upregulated in Keratinocytes during wound repair and early skin carcinogenesis. *J. Invest. Dermatol.*, 124: 849-856. Accessed April 1, 2019. doi: 10.1111/j.0022-202X.2005.23640.x.

Aimes, R. T. and Quigley, J. P. (1995). Matrix metalloproteinase-2 is an interstitial collagenase. Inhibitor-free enzyme catalyzes the cleavage of collagen fibrils and soluble native type I collagen generating the specific 3/4- and 1/4-length fragments. *J. Biol. Chem.*, 270: 5872-5876.

Ashworth, J. L., Murphy, G., Rock, M. J., Sherratt, M. J., Shapiro, S. D., Shuttleworth, C. A. and Kielty, C. M. (1999). Fibrillin degradation by matrix metalloproteinases: implications for connective tissue remodelling. *Biochem. J.*, 340: 171-181.

Bair, E. L., Chen, M. L., McDaniel, K., Sekiguchi, K., Cress, A. E., Nagle, R. B. and Bowden, G. T. (2005). Membrane type 1 matrix metalloprotease cleaves laminin-10 and promotes prostate cancer cell migration. *Neoplasia*, 7: 380-389.

Bar-Or, A., Nuttall, R. K., Duddy, M., Alter, A., Kim, H. J., Ifergan, I., Pennington, C. J., Bourgoin, P., Edwards, D. R. and Yong, V. W. (2003). Analyses of all matrix metalloproteinase members in leukocytes emphasize monocytes as major inflammatory mediators in multiple sclerosis. *Brain*, 126: 2738-2749. Accessed April 1, 2019. doi: 10.1093/brain/awg285.

Bergers, G., Brekken, R., McMahon, G., Vu, T. H., Itoh, T., Tamaki, K., Tanzawa, K., Thorpe, P., Itohara, S., Werb, Z. and Hanahan, D. (2000). Matrix metalloproteinase-9 triggers the angiogenic switch during carcinogenesis. *Nat. Cell. Biol.*, 2: 737-744. Accessed April 1, 2019. doi: 10.1038/35036374.

Birkedal-Hansen, H. (1995). Proteolytic remodeling of extracellular matrix. *Curr. Opin. Cell. Biol.*, 7: 728-735.

Bourd-Boittin, K., Basset, L., Bonnier, D., L'Helgoualc'h, A., Samson, M. and Theret, N. (2009). CX3CL1/fractalkine shedding by human hepatic stellate cells: contribution to chronic inflammation in the liver. *J. Cell. Mol. Med.*, 13: 1526-1535. Accessed April 1, 2019. doi: 10.1111/j.1582-4934.2009.00787.x.

Brandner, J. M., Zorn-Kruppa, M., Yoshida, T., Moll, I., Beck, L. A. and De Benedetto, A. (2015). Epidermal tight junctions in health and disease. *Tissue Barriers*, 3: e974451. Accessed April 1, 2019. doi: 10.4161/21688370.2014.974451.

Braverman, I. M. (2000). The cutaneous microcirculation. *J. Invest. Dermatol. Symp. Proc.*, 5: 3-9. Accessed April 1, 2019. doi: 10.1046/j.1087-0024.2000.00010.x.

Brinckerhoff, C. E., Sirum-Connolly, K. L., Karmilowicz, M. J. and Auble, D. T. (1992). Expression of stromelysin and stromelysin-2 in rabbit and human fibroblasts. *Matrix Suppl.*, 1: 165-175.

Buommino, E., De Filippis, A., Gaudiello, F., Balato, A., Balato, N., Tufano, M. A. and Ayala, F. (2012). Modification of osteopontin and MMP-9 levels in patients with psoriasis on anti-TNF-α therapy. *Arch. Dermatol. Res.*, 304: 481-485.

Butin-Israeli, V., Houser, M. C., Feng, M., Thorp, E. B., Nusrat, A., Parkos, C. A. and Sumagin, R. (2016). Deposition of microparticles by neutrophils onto inflamed epithelium: a new mechanism to disrupt epithelial intercellular adhesions and promote transepithelial migration. *FASEB J.*, 30: 4007-4020. Accessed April 1, 2019. doi: 10.1096/fj.201600734R.

Butler, G. S. and Overall, C. M. (2009). Proteomic identification of multitasking proteins in unexpected locations complicates drug targeting. *Nat. Rev. Drug Discov.*, 8: 935-948.

Cerofolini, L., Fragai, M. and Luchinat, C. (2018). Mechanism and inhibition of matrix metalloproteinases. *Curr. Med. Chem.*, Accessed April 1, 2019. doi: 10.2174/0929867325666180326163523.

Chang, J-H., Javier J. A., Chang G. Y., Oliveira H. B. and Azar, D. T. (2005). Functional characterization of neostatins, the MMP-derived, enzymatic cleavage products of type XVIII collagen. *FEBS Lett.*, 579: 3601-3606. Accessed April 1, 2019. doi: 10.1016/j.febslet.2005.05.043.

Chermnykh, E., Kalabusheva, E. and Vorotelyak, E. (2018). Extracellular matrix as a regulator of epidermal stem cell fate. *Int. J. Mol. Sci.*, 19: pii: E1003. Accessed April 1, 2019. doi: 10.3390/ijms19041003.

Chiricozzi, A., Romanelli P., Volpe E., Borsellino, G. and Romanelli, M. (2018). Scanning the immunopathogenesis of psoriasis. *Int. J. Mol. Sci.*, 19: pii: E179. Accessed April 1, 2019. doi: 10.3390/ijms19010179.

Chiu, P-S and Lai, S-C. (2014). Matrix metalloproteinase-9 leads to blood-brain barrier leakage in mice with eosinophilic meningoencephalitis caused by *Angiostrongylus cantonensis*. *Acta Trop.*, 140: 141-150. Accessed April 1, 2019. doi: 10.1016/j.actatropica.2014.08.015.

Chou, J., Chan, M. F. and Werb, Z. (2016). Metalloproteinases: a functional pathway for myeloid cells. *Microbiol. Spectr.*, 4, Accessed April 1, 2019. doi: 10.1128/microbiolspec.MCHD-0002-2015.

Christophers, E. and Mrowietz. U. (1995). The inflammatory infiltrate in psoriasis. *Clin. Dermatol.*, 13: 131-135.

Cirillo, N., Femiano, F., Gombos, F. and Lanza, A. (2007). Metalloproteinase 9 is the outer executioner of desmoglein 3 in apoptotic keratinocytes. *Oral Dis.*, 13: 341-345. Accessed April 1, 2019. doi: 10.1111/j.1601-0825.2006.01287.x.

Cordiali-Fei, P., Trento, E., D'Agosto, G., Bordignon, V., Mussi, A., Ardigo, M., Mastroianni, A., Vento, A., Solivetti, F., Berardesca, E. and Ensoli, F. (2006). Decreased levels of metalloproteinase-9 and angiogenic factors in skin lesions of patients with psoriatic arthritis after therapy with anti-TNF-α. *J. Autoimmune Dis.*, 3: 5. Accessed April 1, 2019. doi: 10.1186/1740-2557-3-5.

Creamer, D., Allen, M. H., Sousa, A., Poston, R. and Barker, J. N. (1997). Localization of endothelial proliferation and microvascular expansion in active plaque psoriasis. *Br. J. Dermatol.*, 136: 859-865.

Crunkhorn, S. (2017). Ocular disorders: Collagen IV-derived peptide prevents angiogenesis. *Nat. Rev. Drug Discov.*, 16: 166. Accessed April 1, 2019. doi: 10.1038/nrd.2017.38.

Cui, N., Hu, M..and Khalil, R. A. (2017). Biochemical and biological attributes of matrix metalloproteinases. *Prog. Mol. Biol. Transl. Sci.*, 147: 1-73. Accessed April 1, 2019. doi: 10.1016/bs.pmbts.2017.02.005.

d'Ortho, M-P., Will, H., Atkinson, S., Butler, G., Messent, A., Gavrilovic, J., Smith, B., Timpl, R., Zardi, L. and Murphy, G. (1997). Membrane-type matrix metalloproteinases 1 and 2 exhibit broad-spectrum proteolytic capacities comparable to many matrix metalloproteinases. *Eur. J. Biochem.*, 250: 751-757.

Dean, R. A., Butler, G. S., Hamma-Kourbali, Y., Delbe, J., Brigstock, D. R., Courty, J. and Overall, C. M. (2007). Identification of candidate angiogenic inhibitors processed by matrix metalloproteinase 2 (MMP-2) in cell-based proteomic screens: disruption of vascular endothelial growth factor (VEGF)/heparin affin regulatory peptide (pleiotrophin) and VEGF/Connective tissue growth factor angiogenic inhibitory complexes by MMP-2 proteolysis. *Mol. Cell. Biol.*, 27: 8454-8465. Accessed April 1, 2019. doi: 10.1128/mcb.00821-07.

Dean, R. A., Cox, J. H., Bellac, C. L., Doucet, A., Starr, A. E. and Overall, C. M. (2008). Macrophage-specific metalloelastase (MMP-12) truncates and inactivates ELR^+ CXC chemokines and generates CCL2, -7, -8, and -13 antagonists: potential role of the macrophage in terminating polymorphonuclear leukocyte influx. *Blood*, 112: 3455-3464. Accessed April 1, 2019. doi: 10.1182/blood-2007-12-129080.

Debret, R., Antonicelli, F., Theill, A., Hornebeck, W., Bernard, P., Guenounou, M. and Le Naour, R. (2005). Elastin-derived peptides induce a T-helper type 1 polarization of human blood lymphocytes.

Arterioscler. Thromb. Vasc. Biol., 25: 1353-1358. Accessed April 1, 2019. doi: 10.1161/01.ATV.0000168412.50855.9f.

Dopytalska, K., Sobolewski, P., Blaszczak, A., Szymanska, E. and Walecka, I. (2018). Psoriasis in special localizations. *Reumatologia*, 56: 392-398. Accessed April 1, 2019. doi: 10.5114/reum.2018.80718.

Durigova, M., Nagase, H., Mort, J. S., and Roughley, P. J. (2011). MMPs are less efficient than ADAMTS5 in cleaving aggrecan core protein. *Matrix Biol.*, 30: 145-153. Accessed April 1, 2019. doi: 10.1016/ j.matbio.2010.10.007.

Eble, J. A., Ries, A., Lichy, A., Mann, K., Stanton, H., Gavrilovic, J., Murphy, G. and Kuhn, K. (1996). The recognition sites of the integrins α1β1 and α2β1 within collagen IV are protected against gelatinase A attack in the native protein. *J. Biol. Chem.*, 271: 30964-30970.

Endo, K., Takino, T., Miyamori, H., Kinsen, H., Yoshizaki, T., Furukawa, M. and Sato, H. (2003). Cleavage of syndecan-1 by membrane type matrix metalloproteinase-1 stimulates cell migration. *J. Biol. Chem.*, 278: 40764-40770. Accessed April 1, 2019. doi: 10.1074/jbc. M306736200.

Feliciani, C., Vitullo, P., D'Orazi, G., Palmirotta, R., Amerio, P., Pour, S. M., Coscione, G., Amerio, P. L. and Modesti, A. (1997). The 72-kDa and the 92-kDa gelatinases, but not their inhibitors TIMP-1 and TIMP-2, are expressed in early psoriatic lesions. *Exp. Dermatol.*, 6: 321-327.

Ferreras, M., Felbor, U., Lenhard, T., Olsen, B. R. and Delaisse, J. (2000). Generation and degradation of human endostatin proteins by various proteinases. *FEBS Lett.*, 486: 247-251.

Fleischmajer, R., Kuroda, K., Hazan, R., Gordon, R. E., Lebwohl, M. G., Sapadin, A. N., Unda, F., Iehara, N. and Yamada, Y. (2000). Basement membrane alterations in psoriasis are accompanied by epidermal overexpression of MMP-2 and its inhibitor TIMP-2. *J. Invest. Dermatol.*, 115: 771-777. Accessed April 1, 2019. doi: 10.1046/ j.1523-1747.2000.00138.x.

Flisiak, I., Porebski, P., and Chodynicka, B. (2006). Effect of psoriasis activity on metalloproteinase-1 and tissue inhibitor of metalloproteinase-1 in plasma and lesional scales. *Acta Derm.*

Venereol., 86: 17-21. Accessed April 1, 2019. doi: 10.1080/ 00015550510011600.

Fosang, A. J., Last, K., Knauper, V., Murphy, G. and Neame, P. J. (1996). Degradation of cartilage aggrecan by collagenase-3 (MMP-13). *FEBS Lett.*, 380: 17-20.

Fosang, A. J., Last, K., Knauper, V., Neame, P. J., Murphy, G., Hardingham, T. E., Tschesche, H. and Hamilton, J. A. (1993). Fibroblast and neutrophil collagenases cleave at two sites in the cartilage aggrecan interglobular domain. *Biochem. J.*, 295: 273-276.

Fosang, A. J., Last, K., Neame, P. J., Murphy, G., Knauper, V., Tschesche, H., Hughes, C. E., Caterson, B. and Hardingham, T. E. (1994). Neutrophil collagenase (MMP-8) cleaves at the aggrecanase site E373-A374 in the interglobular domain of cartilage aggrecan. *Biochem. J.*, 304: 347-351.

Gearing, A. J., Beckett, P., Christodoulou, M., Churchill, M., Clements, J., Davidson, A. H., Drummond, A. H., Galloway, W. A., Gilbert, R., Gordon, J. L. et al. (1994). Processing of tumour necrosis factor-α precursor by metalloproteinases. *Nature*, 370: 555-557. Accessed April 1, 2019. doi: 10.1038/370555a0.

Gioia, M., Monaco, S., Fasciglione, G. F., Coletti, A., Modesti, A., Marini, S. and Coletta, M. (2007). Characterization of the mechanisms by which gelatinase A, neutrophil collagenase, and membrane-type metalloproteinase MMP-14 recognize collagen I and enzymatically process the two α-chains. *J. Mol. Biol.*, 368: 1101-1113. Accessed April 1, 2019. doi: 10.1016/j.jmb.2007.02.076.

Gronski, T. J., Jr., Martin, R. L., Kobayashi, D. K., Walsh, B. C., Holman, M. C., Huber, M., Van Wart, H. E. and Shapiro, S. D. (1997). Hydrolysis of a broad spectrum of extracellular matrix proteins by human macrophage elastase. *J. Biol. Chem.*, 272: 12189-12194.

Gubbiotti, M. A., Neill, Thomas and Iozzo, Renato V. (2017). A current view of perlecan in physiology and pathology: A mosaic of functions. *Matrix Biol.*, 57-58: 285-298. Accessed April 1, 2019. doi: 10.1016/j.matbio.2016.09.003.

Halpert, I., Sires, U. I., Roby, J. D., Potter-Perigo, S., Wight, T. N., Shapiro, S. D., Welgus, H. G., Wickline, S. A. and Parks, W. C. (1996). Matrilysin is expressed by lipid-laden macrophages at sites of potential rupture in atherosclerotic lesions and localizes to areas of versican deposition, a proteoglycan substrate for the enzyme. *Proc. Natl. Acad. Sci. U. S. A.*, 93: 9748-9753.

Heinz, A., Jung, M. C., Duca, L., Sippl, W., Taddese, S., Ihling, C., Rusciani, A., Jahreis, G., Weiss, A. S., Neubert, R. H. and Schmelzer, C. E. (2010). Degradation of tropoelastin by matrix metalloproteinases – cleavage site specificities and release of matrikines. *FEBS J.*, 277: 1939-1956. Accessed April 1, 2019. doi: 10.1111/j.1742-4658.2010.07616.x.

Hern, S., Stanton, A. W., Mellor, R., Levick, J. R. and Mortimer, P. S. (1999). Control of cutaneous blood vessels in psoriatic plaques. *J. Invest. Dermatol.*, 113: 127-132. Accessed April 1, 2019. doi: 10.1046/j.1523-1747.1999.00638.x.

Hitchon, C. A., Danning, C. L., Illei, G. G., El-Gabalawy, H. S. and Boumpas, D. T. (2002). Gelatinase expression and activity in the synovium and skin of patients with erosive psoriatic arthritis. *J. Rheumatol.*, 29: 107-117.

Houghton, A. M., Quintero, P. A., Perkins, D. L., Kobayashi, D. K., Kelley, D. G., Marconcini, L. A., Mecham, R. P., Senior, R. M. and Shapiro, S. D. (2006). Elastin fragments drive disease progression in a murine model of emphysema. *J. Clin. Invest.*, 116: 753-759. Accessed April 1, 2019. doi: 10.1172/jci25617.

Imai, K., Shikata, Hideo and Okada, Yasunori. (1995). Degradation of vitronectin by matrix metalloproteinases-1, -2, -3, -7 and -9. *FEBS Lett.*, 369: 249-251.

Impola, U., Toriseva, M., Suomela, S., Jeskanen, L., Hieta, N., Jahkola, T., Grenman, R., Kahari, V. M. and Saarialho-Kere, U. (2003). Matrix metalloproteinase-19 is expressed by proliferating epithelium but disappears with neoplastic dedifferentiation. *Int. J. Cancer*, 103: 709-716. Accessed April 1, 2019. doi: 10.1002/ijc.10902.

Ito, A., Mukaiyama, A., Itoh, Y., Nagase, H., Thogersen, I. B., Enghild, J. J., Sasaguri, Y. and Mori, Y. (1996). Degradation of interleukin 1β by matrix metalloproteinases. *J. Biol. Chem.*, 271: 14657-14660.

Jiao, Y., Feng, Y., Zhan, Y., Wang, R., Zheng, S., Liu, W. and Zeng, X. (2012). Matrix metalloproteinase-2 promotes αvβ3 integrin-mediated adhesion and migration of human melanoma cells by cleaving fibronectin. *PLoS One*, 7: e41591. Accessed April 1, 2019. doi: 10.1371/journal.pone.0041591.

Kajita, M., Itoh, Y., Chiba, T., Mori, H., Okada, A., Kinoh, H. and Seiki, M. (2001). Membrane-type 1 matrix metalloproteinase cleaves CD44 and promotes cell migration. *J. Cell. Biol.*, 153: 893-904.

Kang, T., Yi, J., Guo, A., Wang, X., Overall, C. M., Jiang, W., Elde, R., Borregaard, N. and Pei, D. (2001). Subcellular distribution and cytokine- and chemokine-regulated secretion of leukolysin/MT6-MMP/MMP-25 in neutrophils. *J. Biol. Chem.*, 276: 21960-21968. Accessed April 1, 2019. doi: 10.1074/jbc.M007997200.

Kerkela, E., Ala-aho, R., Lohi, J., Grenman, R., Kahari, M. and Saarialho-Kere, U. (2001). Differential patterns of stromelysin-2 (MMP-10) and MT1-MMP (MMP-14) expression in epithelial skin cancers. *Br. J. Cancer*, 84: 659-669. Accessed April 1, 2019. doi: 10.1054/bjoc.2000.1634.

Khamaganova, I. V., Almazova, A. A., Lebedeva, G. A. and Ermachenko, A. V. (2015). Problems in the epidemiology of psoriasis. *Clinical Dermatology and Venerology*, no. 1: 12-16.

Kirschner, N., Poetzl, C., von den Driesch, P., Wladykowski, E., Moll, I., Behne, M. J. and Brandner, J. M. (2009). Alteration of tight junction proteins is an early event in psoriasis: putative involvement of proinflammatory cytokines. *Am. J. Pathol.*, 175: 1095-1106. Accessed April 1, 2019. doi: 10.2353/ajpath.2009.080973.

Knauper, V., Cowell, S., Smith, B., Lopez-Otin, C., O'Shea, M., Morris, H., Zardi, L. and Murphy, G. (1997). The role of the C-terminal domain of human collagenase-3 (MMP-13) in the activation of procollagenase-3, substrate specificity, and tissue inhibitor of metalloproteinase interaction. *J. Biol. Chem.*, 272: 7608-7616.

Korff, T. and Augustin, H. G. (1999). Tensional forces in fibrillar extracellular matrices control directional capillary sprouting. *J. Cell. Sci.*, 112: 3249-3258.

Kuivanen, T., Ahokas, K., Virolainen, S., Jahkola, T., Holtta, E., Saksela, O. and Saarialho-Kere, U. (2005). MMP-21 is upregulated at early stages of melanoma progression but disappears with more aggressive phenotype. *Virchows Arch.*, 447: 954-960. Accessed April 1, 2019. doi: 10.1007/s00428-005-0046-8.

Lebwohl, M. (2018). Psoriasis. *Ann. Intern. Med.*, 168: ITC49-ITC64. Accessed April 1, 2019. doi: 10.7326/aitc201804030.

Lee, S., Jilani, S. M., Nikolova, G. V., Carpizo, D. and Iruela-Arispe, M. L. (2005). Processing of VEGF-A by matrix metalloproteinases regulates bioavailability and vascular patterning in tumors. *J. Cell. Biol.*, 169: 681-691. Accessed April 1, 2019. doi: 10.1083/jcb.200409115.

Li, C., Lasse, S., Lee, P., Nakasaki, M., Chen, S. W., Yamasaki, K., Gallo, R. L. and Jamora, C. (2010). Development of atopic dermatitis-like skin disease from the chronic loss of epidermal caspase-8. *Proc. Natl. Acad. Sci. U. S. A.*, 107: 22249-22254. Accessed April 1, 2019. doi: 10.1073/pnas.1009751108.

Li, Q, Park, P. W., Wilson, C. L. and Parks, W. C. (2002). Matrilysin shedding of syndecan-1 regulates chemokine mobilization and transepithelial efflux of neutrophils in acute lung injury. *Cell*, 111: 635-646.

Lin, H-C, Chang, J. H., Jain, S., Gabison, E. E., Kure, T., Kato, T., Fukai, N. and Azar, D. T. (2001). Matrilysin cleavage of corneal collagen type XVIII NC1 domain and generation of a 28-kDa fragment. *Invest. Ophthalmol. Vis. Sci.*, 42: 2517-2524.

Lindqvist, U., Phil-Lundin, I. and Engstrom-Laurent, A. (2012). Dermal distribution of hyaluronan in psoriatic arthritis; coexistence of CD44, MMP3 and MMP9. *Acta Derm. Venereol.*, 92: 372-377. Accessed April 1, 2019. doi: 10.2340/00015555-1286.

Madlener, M., Mauch, C., Conca, W., Brauchle, M., Parks, W. C. and Werner, S. (1996). Regulation of the expression of stromelysin-2 by growth factors in keratinocytes: implications for normal and impaired wound healing. *Biochem. J.*, 320: 659-664.

Maeda, S., Dean, D. D., Gomez, R., Schwartz, Z. and Boyan, B. D. (2002). The first stage of transforming growth factor β1 activation is release of the large latent complex from the extracellular matrix of growth plate chondrocytes by matrix vesicle stromelysin-1 (MMP-3). *Calcif. Tissue Int.*, 70: 54-65. Accessed April 1, 2019. doi: 10.1007/s002230010032.

Marchenko, G. N., Ratnikov, B. I., Rozanov, D. V., Godzik, A., Deryugina, E. I. and Strongin, A. Y. (2001). Characterization of matrix metalloproteinase-26, a novel metalloproteinase widely expressed in cancer cells of epithelial origin. *Biochem. J.*, 356: 705-718.

Mayer, U., Mann, K., Timpl, R. and Murphy, G. (1993). Sites of nidogen cleavage by proteases involved in tissue homeostasis and remodelling. *Eur. J. Biochem.*, 217: 877-884.

McQuibban, A., Gong, J. H., Tam, E. M., McCulloch, C. A., Clark-Lewis, I. and Overall, C. M. (2000). Inflammation dampened by gelatinase A cleavage of monocyte chemoattractant protein-3. *Science*, 289: 1202-1206.

McQuibban, A., Gong, J. H., Wong, J. P., Wallace, J. L., Clark-Lewis, I. and Overall, C. M. (2002). Matrix metalloproteinase processing of monocyte chemoattractant proteins generates CC chemokine receptor antagonists with anti-inflammatory properties in vivo. *Blood*, 100: 1160-1167.

Meerovitch, K., Bergeron, F., Leblond, L., Grouix, B., Poirier, C., Bubenik, M., Chan, L., Gourdeau, H., Bowlin, T. and Attardo, G. (2003). A novel RGD antagonist that targets both αvβ3 and α5β1 induces apoptosis of angiogenic endothelial cells on type I collagen. *Vascul. Pharmacol.*, 40: 77-89.

Mezentsev, A., Nikolaev, A. and Bruskin, S. (2014). Matrix metallo-proteinases and their role in psoriasis. *Gene*, 540: 1-10. Accessed April 1, 2019. doi: 10.1016/j.gene.2014.01.068.

Micali, G., Lacarrubba, F., Musumeci, M. L., Massimino, D. and Nasca, M. R. (2010). Cutaneous vascular patterns in psoriasis. *Int. J. Dermatol.*,

49: 249-256. Accessed April 1, 2019. doi: 10.1111/j.1365-4632.2009.04287.x.

Mitsui, H., Suarez-Farinas, M., Belkin, D. A., Levenkova, N., Fuentes-Duculan, J., Coats, I., Fujita, H. and Krueger, J. G. (2012). Combined use of laser capture microdissection and cDNA microarray analysis identifies locally expressed disease-related genes in focal regions of psoriasis vulgaris skin lesions. *J. Invest. Dermatol.*, 132: 1615-1626. Accessed April 1, 2019. doi: 10.1038/jid.2012.33.

Mochizuki, S., Brassart, B. and Hinek, A. (2002). Signaling pathways transduced through the elastin receptor facilitate proliferation of arterial smooth muscle cells. *J. Biol. Chem.*, 277: 44854-44863. Accessed April 1, 2019. doi: 10.1074/jbc.M205630200.

Mydel, P., Shipley, J. M., Adair-Kirk, T. L., Kelley, D. G., Broekelmann, T. J., Mecham, R. P. and Senior, R. M. (2008). Neutrophil elastase cleaves laminin-332 (laminin-5) generating peptides that are chemotactic for neutrophils. *J. Biol. Chem.*, 283: 9513-9522. Accessed April 1, 2019. doi: 10.1074/jbc.M706239200.

Nissinen, L. and Kahari, V-M. M. (2014). Matrix metalloproteinases in inflammation. *Biochim. Biophys. Acta*, 1840: 2571-2580. Accessed April 1, 2019. doi: 10.1016/j.bbagen.2014.03.007.

Noe, V., Fingleton, B., Jacobs, K., Crawford, H. C., Vermeulen, S., Steelant, W., Bruyneel, E., Matrisian, L. M. and Mareel, M. (2001). Release of an invasion promoter E-cadherin fragment by matrilysin and stromelysin-1. *J. Cell. Sci.*, 114: 111-118.

Ntayi, C., Labrousse, A. L., Debret, R., Birembaut, P., Bellon, G., Antonicelli, F., Hornebeck, W. and Bernard, P. (2004). Elastin-derived peptides upregulate matrix metalloproteinase-2-mediated melanoma cell invasion through elastin-binding protein. *J. Invest. Dermatol.*, 122: 256-265. Accessed April 1, 2019. doi: 10.1046/j.0022-202X.2004.22228.x.

Ohuchi, E., Imai, K., Fujii, Y., Sato, H., Seiki, M. and Okada, Y. (1997). Membrane type 1 matrix metalloproteinase digests interstitial collagens and other extracellular matrix macromolecules. *J. Biol. Chem.*, 272: 2446-2451.

Okada, Y., Naka, K., Kawamura, K., Matsumoto, T., Nakanishi, I., Fujimoto, N., Sato, H. and Seiki, M. (1995). Localization of matrix metalloproteinase 9 (92-kilodalton gelatinase/type IV collagenase = gelatinase B) in osteoclasts: implications for bone resorption. *Lab. Invest.*, 72: 311-322.

Oku, N., Sasabe, E., Ueta, E., Yamamoto, T. and Osaki. T. (2006). Tight junction protein claudin-1 enhances the invasive activity of oral squamous cell carcinoma cells by promoting cleavage of laminin-5 γ2 chain via matrix metalloproteinase (MMP)-2 and membrane-type MMP-1. *Cancer Res.*, 66: 5251-5257. Accessed April 1, 2019. doi: 10.1158/0008-5472.can-05-4478.

Pearce, W. H., and Shively, Vera P. (2006). Abdominal aortic aneurysm as a complex multifactorial disease: interactions of polymorphisms of inflammatory genes, features of autoimmunity, and current status of MMPs. *Ann. N. Y. Acad. Sci.*, 1085: 117-132. Accessed April 1, 2019. doi: 10.1196/annals.1383.025.

Pickart, L, and Margolina, A. (2018). Regenerative and Protective Actions of the GHK-Cu Peptide in the Light of the New Gene Data. *Int. J. Mol. Sci.*, 19: pii: E1987. Accessed April 1, 2019. doi: 10.3390/ ijms19071987.

Pilcher, B. K., Dumin, J. A., Sudbeck, B. D., Krane, S. M., Welgus, H. G. and Parks, W. C. (1997). The activity of collagenase-1 is required for keratinocyte migration on a type I collagen matrix. *J. Cell. Biol.*, 137: 1445-1457.

Poole, T. J., and Zetter, B. R. (1983). Stimulation of rat peritoneal mast cell migration by tumor-derived peptides. *Cancer Res.*, 43: 5857-5861.

Ramakrishnan, S., Nguyen, T. M., Subramanian, I. V. and Kelekar, A. (2007). Autophagy and angiogenesis inhibition. *Autophagy*, 3: 512-515.

Reno, F., Sabbatini, M., Stella, M., Magliacani, G. and Cannas, M. (2005). Effect of in vitro mechanical compression on Epilysin (matrix metalloproteinase-28) expression in hypertrophic scars. *Wound Repair Regen.*, 13: 255-261. Accessed April 1, 2019. doi: 10.1111/j.1067-1927.2005.130307.x.

Ricard-Blum, S., and Vallet, S. D. (2017). Fragments generated upon extracellular matrix remodeling: Biological regulators and potential drugs. *Matrix Biol.*, 300-313. Accessed April 1, 2019. doi: 10.1016/j.matbio.2017.11.005.

Rims, C. R., and McGuire, J. K. (2014). Matrilysin (MMP-7) catalytic activity regulates β-catenin localization and signaling activation in lung epithelial cells. *Exp. Lung Res.*, 40: 126-136. Accessed April 1, 2019. doi: 10.3109/01902148.2014.890681.

Ritchlin, C. T, Colbert, R. A. and Gladman, D. D. (2017). Psoriatic arthritis. *N. Engl. J. Med.*, 376: 2095-2096. Accessed April 1, 2019. doi: 10.1056/NEJMra1505557.

Saalbach, A., Arnhold, J., Lessig, J., Simon, J. C., and Anderegg, U. (2008). Human Thy-1 induces secretion of matrix metalloproteinase-9 and CXCL8 from human neutrophils. *Eur. J. Immunol.*, 38: 1391-1403. Accessed April 1, 2019. doi: 10.1002/eji.200737901.

Saarialho-Kere, U. K., Pentland, A. P., Birkedal-Hansen, H., Parks, W. C. and Welgus, H. G. (1994). Distinct populations of basal keratinocytes express stromelysin-1 and stromelysin-2 in chronic wounds. *J. Clin. Invest.*, 94: 79-88. Accessed April 1, 2019. doi: 10.1172/jci117351.

Saarialho-Kere, U., Kerkela, E., Jahkola, T., Suomela, S., Keski-Oja, J. and Lohi, J. (2002). Epilysin (MMP-28) expression is associated with cell proliferation during epithelial repair. *J. Invest. Dermatol.*, 119: 14-21. Accessed April 1, 2019. doi: 10.1046/j.1523-1747.2002.01790.x.

Sadowski, T., Dietrich, S., Muller, M., Havlickova, B., Schunck, M., Proksch, E., Muller, M. S. and Sedlacek, R. (2003). Matrix metalloproteinase-19 expression in normal and diseased skin: dysregulation by epidermal proliferation. *J. Invest. Dermatol.*, 121: 989-996. Accessed April 1, 2019. doi: 10.1046/j.1523-1747.2003.12526.x.

Sasaki, T., Mann, K., Murphy, G., Chu, M. L. and Timpl, R. (1996). Different susceptibilities of fibulin-1 and fibulin-2 to cleavage by matrix metalloproteinases and other tissue proteases. *Eur. J. Biochem.*, 240: 427-434.

Schonherr, E. and Hausser, H.-J. (2000). Extracellular matrix and cytokines: a functional unit. *Dev. Immunol.*, 7: 89-101.

Sedlacek, R., Mauch, S., Kolb, B., Schatzlein, C., Eibel, H., Peter, H. H., Schmitt, J. and Krawinkel, U. (1998). Matrix metalloproteinase MMP-19 (RASI-1) is expressed on the surface of activated peripheral blood mononuclear cells and is detected as an autoantigen in rheumatoid arthritis. *Immunobiology*, 198: 408-423. Accessed April 1, 2019. doi: 10.1016/s0171-2985(98)80049-1.

Seltzer, Jo L., Eisen, A. Z., Bauer, E. A., Morris, N. P., Glanville, R. W. and Burgeson, R. E. (1989). Cleavage of type VII collagen by interstitial collagenase and type IV collagenase (gelatinase) derived from human skin. *J. Biol. Chem.*, 264: 3822-3826.

Senior, R. M., Griffin, G. L., Fliszar, C. J., Shapiro, S. D., Goldberg, G. I. and Welgus, H. G. (1991). Human 92- and 72-kilodalton type IV collagenases are elastases. *J. Biol. Chem.*, 266: 7870-7875.

Senior, R. M., Griffin, G. L., Mecham, R. P., Wrenn, D. S., Prasad, K. U. and Urry, D. W. (1984). Val-Gly-Val-Ala-Pro-Gly, a repeating peptide in elastin, is chemotactic for fibroblasts and monocytes. *J. Cell. Biol.*, 99: 870-874.

Siefert, S. A., and Sarkar, Rajabrata. (2012). Matrix metalloproteinases in vascular physiology and disease. *Vascular*, 20: 210-216. Accessed April 1, 2019. doi: 10.1258/vasc.2011.201202.

Simonetti, O., Lucarini, G., Goteri, G., Zizzi, A., Biagini, G., Lo Muzio, L. and Offidani, A. (2006). VEGF is likely a key factor in the link between inflammation and angiogenesis in psoriasis: results of an immunohistochemical study. *Int. J. Immunopathol. Pharmacol.*, 19: 751-760. Accessed April 1, 2019. doi: 10.1177/039463200601900405.

Siri, A., Knauper, V., Veirana, N., Caocci, F., Murphy, G. and Zardi, L. (1995). Different susceptibility of small and large human tenascin-C isoforms to degradation by matrix metalloproteinases. *J. Biol. Chem.*, 270: 8650-8654.

Stahle-Backdahl, M., Inoue, M., Guidice, G. J. and Parks, W. C. (1994). 92-kD gelatinase is produced by eosinophils at the site of blister formation in bullous pemphigoid and cleaves the extracellular domain of recombinant 180-kD bullous pemphigoid autoantigen. *J. Clin. Invest.*, 93: 2022-2030. Accessed April 1, 2019. doi: 10.1172/jci117196.

Starodubtseva, N. L., Sobolev, V. V., Soboleva, A. G., Nikolaev, A. A. and Bruskin. S. A. (2011). Expression of genes for metalloproteinases (MMP-1, MMP-2, MMP-9, and MMP-12) associated with psoriasis. *Genetika*, 47: 1254-1261.

Starr, A. E., Dufour A., Maier J., and Overall, C. M. (2012). Biochemical analysis of matrix metalloproteinase activation of chemokines CCL15 and CCL23 and increased glycosaminoglycan binding of CCL16. *J. Biol. Chem.*, 287: 5848-5860. Accessed April 1, 2019. doi: 10.1074/jbc.M111.314609.

Sternlicht, M. D., and Werb. Z. (2001). How matrix metalloproteinases regulate cell behavior. *Annu. Rev. Cell. Dev. Biol.*, 17: 463-516. Accessed April 1, 2019. doi: 10.1146/annurev.cellbio.17.1.463.

Stinco, G., Lautieri, S. and Patrone, P. (2006). Video capillaroscopic study in psoriatic patients treated with tacalcitol. *G. Ital. Dermatol. Venereol.*, 141: 227-231.

Stracke, J. O., Fosang, A. J., Last, K., Mercuri, F. A., Pendas, A. M., Llano, E., Perris, R., Di Cesare, P. E., Murphy, G. and Knauper, V. (2000). Matrix metalloproteinases 19 and 20 cleave aggrecan and cartilage oligomeric matrix protein (COMP). *FEBS Lett.*, 478: 52-56.

Stracke, J. O., Hutton, M., Stewart, M., Pendas, A. M., Smith, B., Lopez-Otin, C., Murphy, G. and Knauper, V. (2000). Biochemical characterization of the catalytic domain of human matrix metalloproteinase 19. Evidence for a role as a potent basement membrane degrading enzyme. *J. Biol. Chem.*, 275: 14809-14816.

Suomela, S., Kariniemi, A. L., Impola, U., Karvonen, S. L., Snellman, E., Uurasmaa, T., Peltonen, J. and Saarialho-Kere, U. (2003). Matrix metalloproteinase-19 is expressed by keratinocytes in psoriasis. *Acta Derm. Venereol.*, 83: 108-114.

Suomela, S., Kariniemi, A. L., Snellman, E. and Saarialho-Kere, U. (2001). Metalloelastase (MMP-12) and 92-kDa gelatinase (MMP-9) as well as their inhibitors, TIMP-1 and -3, are expressed in psoriatic lesions. *Exp. Dermatol.*, 10: 175-183.

Suzuki, M., Raab, G., Moses, M. A., Fernandez, C. A. and Klagsbrun, M. (1997). Matrix metalloproteinase-3 releases active heparin-binding

EGF-like growth factor by cleavage at a specific juxtamembrane site. *J. Biol. Chem.*, 272: 31730-31737.

Takeshita, J., Grewal, S., Langan, S. M., Mehta, N. N., Ogdie, A., Van Voorhees, A. S. and Gelfand, J. M. (2017). Psoriasis and comorbid diseases: Epidemiology. *J. Am. Acad. Dermatol.*, 76: 377-390. Accessed April 1, 2019. doi: 10.1016/j.jaad.2016.07.064.

Tsaousi, A., Witte, E., Witte, K., Rowert-Huber, H. J., Volk, H. D., Sterry, W., Wolk, K., Schneider-Burrus, S. and Sabat, R. (2016). MMP8 is increased in lesions and blood of acne inversa patients: apotential link to skin destruction and metabolic alterations. *Mediators Inflamm.*, 4097574. Accessed April 1, 2019. doi: 10.1155/2016/4097574.

Uhlen, M., Fagerberg, L., Hallstrom, B. M., Lindskog, C., Oksvold, P., Mardinoglu, A., Sivertsson, A., Kampf, C., Sjostedt, E., Asplund, A., Olsson, I., Edlund, K., Lundberg, E. Navani, S., Szigyarto, C. A., Odeberg, J., Djureinovic, D., Takanen, J. O., Hober, S., Alm, T., Edqvist, P. H., Berling, H., Tegel, H., Mulder, J., Rockberg, J., Nilsson, P., Schwenk, J. M., Hamsten, M., von Feilitzen, K., Forsberg, M., Persson, L., Johansson, F., Zwahlen, M., von Heijne, G., Nielsen, J. and Ponten, F. (2015). Proteomics. tissue-based map of the human proteome. *Science*, 347: 1260419. Accessed April 1, 2019. doi: 10.1126/science.1260419.

Van Den Steen P. E, Wuyts A, Husson S. J, Proost P, Van Damme J. and Opdenakker G. (2003). Gelatinase B/MMP-9 and neutrophil collagenase/MMP-8 process the chemokines human GCP-2/CXCL6, ENA-78/CXCL5 and mouse GCP-2/LIX and modulate their physiological activities. *Eur. J. Biochem.* 270: 3739-3749.

Van Wart, H. E. (1992). Human neutrophil collagenase. *Matrix Suppl.*, 1: 31-36.

Watson, R. E., Poddar, R., Walker, J. M., McGuill, I., Hoare, L. M., Griffiths, C. E. and O'Neill C. A. (2007). Altered claudin expression is a feature of chronic plaque psoriasis. *J. Pathol.*, 212: 450-458. Accessed April 1, 2019. doi: 10.1002/path.2200.

Weathington, N. M., van Houwelingen, A. H., Noerager, B. D., Jackson, P. L., Kraneveld, A. D., Galin, F. S., Folkerts, G., Nijkamp, F. P. and

Blalock, J. E. (2006). A novel peptide CXCR ligand derived from extracellular matrix degradation during airway inflammation. *Nat. Med.*, 12: 317-323. Accessed April 1, 2019. doi: 10.1038/nm1361.

Wolk, K., Mitsui, H., Witte, K., Gellrich, S., Gulati, N., Humme, D., Witte, E., Gonsior, M., Beyer, M., Kadin, M. E., Volk, H. D., Krueger, J. G., Sterry, W. and Sabat, R. (2014). Deficient cutaneous antibacterial competence in cutaneous T-cell lymphomas: role of T_{h2}-mediated biased T_{h17} function. *Clin. Cancer Res.*, 20: 5507-5516. Accessed April 1, 2019. doi: 10.1158/1078-0432.ccr-14-0707.

Yamamoto, K., Miyazaki, K. and Higashi, S. (2010). Cholesterol sulfate alters substrate preference of matrix metalloproteinase-7 and promotes degradations of pericellular laminin-332 and fibronectin. *J. Biol. Chem.*, 285: 28862-28873. Accessed April 1, 2019. doi: 10.1074/jbc.M110. 136994.

Yu, Q., and Stamenkovic, I. (2000). Cell surface-localized matrix metalloproteinase-9 proteolytically activates TGF-β and promotes tumor invasion and angiogenesis. *Genes Dev.*, 14: 163-176.

Zetter, B, R., Rasmussen, N. and Brown, L. (1985). An in vivo assay for chemoattractant activity. *Lab. Invest.*, 53: 362-368.

Zhang, L, Yang, M., Yang, D., Cavey, G., Davidson, P., and Gibson, G. (2010). Molecular interactions of MMP-13 C-terminal domain with chondrocyte proteins. *Connect. Tissue Res.*, 51: 230-239. Accessed April 1, 2019. doi: 10.3109/03008200903288902.

Zhang, X, Chen, C. T., Bhargava, M. and Torzilli, P. A. (2012). A comparative study of fibronectin cleavage by MMP-1, -3, -13, and -14. *Cartilage*, 3: 267-277. Accessed April 1, 2019. doi: 10.1177/ 1947603511435273.

Zhao, X, Sun, B., Li, Y., Liu, Y., Zhang, D., Wang, X., Gu, Q., Zhao, J., Dong, X., Liu, Z. and Che, N. (2015). Dual effects of collagenase-3 on melanoma: metastasis promotion and disruption of vasculogenic mimicry. *Oncotarget*, 6: 8890-8899. Accessed April 1, 2019. doi: 10.18632/oncotarget.3189.

In: A Closer Look at Metalloproteinases
Editor: Lena Goodwin
ISBN: 978-1-53616-517-3

Chapter 5

PHYSIOLOGICAL ROLE OF METALLOPROTEINASES DURING PREGNANCY

Bruno Zavan*[*]*, Évila da Silva Lopes Salles, Renato de Oliveira Horvath, Andrea do Amarante-Paffaro and Valdemar Antonio Paffaro Junior
Biomedical Science Institute, UNIFAL, Alfenas. Brazil

ABSTRACT

The degradation of proteins and connective tissue from extracellular matrix can be a crucial event for tissues in intense remodeling process, with proteases as mediators called matrix metalloproteinases (MMPs). Pregnancy is a very peculiar event in which the uterus undergoes intense morphophysiological modifications, especially related to vascular remodeling and angiogenesis, as well as trophoblast invasion, which occurs in early pregnancy period until reaching maternal spiral arteries and replacing the endothelium forming the endovascular-trophoblast. Such vascular modifications ensure the high-flow and low-pressure blood supply

[*] Corresponding Author's E-mail: bruno_zavan@yahoo.com.br.

to the developing fetus, which in turn requires more and more space, causing the uterus to become enlarged and distended. In this context, an orchestrated regulation of MMPs and their endogenous inhibitors - such as tissue inhibitors of metalloproteinases (TIMP) - are essential for proper gestational development. The disordered uteroplacental remodeling is associated with several obstetric complications, such as placenta accreta, fetal growth restriction, abortion, preterm delivery and pre-eclampsia. Studies of MMPs involved in the uterine remodeling process and their relationship to local maternal immune cells, as well as fetal trophoblastic cells, still represent a very promising field of research, since the elucidation of the involved processes can boost the development of strategies to prevent complications in woman's health, obstetrics and for consequences in offspring adulthood.

Keywords: pregnancy, extracellular matrix, uterine remodeling

INTRODUCTION

The cells in a tissue are immersed in an environment consisting of multifunctional macromolecules called extracellular matrix. With multiples domains, the extracellular matrix can interact with other surrounding components, as well as with growth factors or cell surface receptors, such as integrins, which in turn are connected to the cytoskeleton and also interact with complex signaling networks that regulate gene expression and, finally, cellular activities. Thus, the extracellular matrix is not limited to its structural function, but becomes fundamental for the processes of cell proliferation, differentiation, and survival [1-3].

During pregnancy, the uterus undergoes intense morphological modifications including the implantation process, followed by decidualization, trophoblast invasion, vascular remodeling, uterine growth, tissue adaptation for parturition, and finally, the uterus initiates a postpartum involution. In this context, the degradation of proteins and connective tissue from extracellular matrix is crucial to guarantee tissue plasticity and, consequently, the healthy progress of pregnancy.

The degradation and remodeling of proteins and connective tissue from extracellular matrix are events mediated by Matrix Metalloproteinases

(MMPs), that are enzymes from a proteolytic zinc-dependent family characterized by their structure and substrate specificities into collagenases (MMP-1, MMP-8 and MMP-13), gelatinases (MMP-2 and 9), stromelysins (MMP-3, MMP-10 and MMP-11), matrilysins (MMP-7), metalloelastase (MMP-12), and membrane type-MMPs [4].

To ensure the precise tissue remodeling, MMPs need to be properly regulated by specific endogenous tissue inhibitors of metalloproteinases (TIMPs), which comprise a family of 4 protease inhibitors: TIMP-1, TIMP-2, TIMP-3, and TIMP-4. Both MMPs and TIMPs are secreted by a variety of connective tissues and pro-inflammatory cells, including fibroblasts, osteoblasts, endothelial cells, macrophages, neutrophils, and lymphocytes [4].

Implantation

When a successful mating occurs, the fertilized oocyte develops to morula stage while moving in the uterine tube lumen until it reaches the uterine cavity where the implantation may take place. In mice, the implantation occurs between gestations day 4-5 [5] whereas in humans this occurs at the end of the week 1 [6], and this event can be described as a process starting with the apposition and adhesion of a competent blastocyst to the apical surface of the endometrial epithelium - process also called as "attachment" [7]. The attachment is only possible if the uterine luminal epithelium is receptive, which in turn occurs just in a short period of time termed "implantation window" [8].

During blastocyst implantation, the trophoblast cells rapidly penetrate at endometrium, and this process can be considered similar to the invasion of basement membranes and stroma by tumor cells [9]. The cancer cells migrate into tissues and vessels probably involving a series of cell-matrix interactions. MMPs and TIMPs are thought to be key mediators for extracellular matrix reorganization during implantation process, emphasizing the need for a thin balance between such factors for the accuracy of the event [10-15]. Specifically for the implantation process, the

MMP-1 and MMP-3 appear to be fundamental since they are expressed by the extravillous trophoblast in this period [16-17] and their altered levels are associated with infertility and recurrent miscarriages [18].

In addition to the role on extracellular matrix degradation by MMPs in the implantation period, they also participate on bioactivity of growth factor, cytokines and angiogenic factors by making them biologically active after proteolytic cleavage [19-21].

Vascular Remodeling and Decidualization

The process of angiogenesis and decidualization takes place shortly after blastocyst implantation concurrently with the influx of immune cells into the uterine environment. There are many studies pointing out the importance of these cells in the production of angiogenic factors and MMPs and, therefore, it is characterized as an essential process for a healthy gestation.

The decidual immune cells comprises almost 40% of total cells during pregnancy [22]. These cells are present abundantly, as they play two important roles in gestation: (i) protecting the mother from blood borne pathogens, or transmitted through the mucosa; (ii) and at the same time, these cells support the development of the hemiallogenic conceptus [23]. In this way, it is important to emphasize that the placenta is not immunologically silent, which implies that the interaction of maternal immune cells with fetal cells must be in perfect harmony for gestational success [24].

Uterine Natural Killer (uNK) cells stand out among other immune cells (macrophages, dendritic cells and T cells), not only by reaching expressive quantitative numbers (70% of the early decidual leukocytes) [22, 25, 26], but also by coupling reduced cytotoxicity with an ability to secrete cytokines and angiogenic factors [27]. Studies have shown that uNK cells (also known as decidual or dNK by many authors) are able to produce Vascular Endothelial Growth Factor (VEGF) A, VEGF-C, Placental Growth Factor (PGF), Angiopoietin-1 and -2, MMP-2, transforming growth factor beta 1 (TGFb1) in humans [28-31], and VEGFA, VEGFC, PGF, delta-like ligand

1 (DLL1), TGFb1, MMPs, tumor necrosis factor A, and inducible nitric oxide synthase (iNOS) in mice [32-38].

The uNK cells express molecules related to degradation of the extracellular matrix, such as MMP-2, MMP-9, TIMP-1, TIMP-2 and TIMP-3; and they are implicated in the promotion of spiral arteries remodeling and the regulation of extravillous trophoblast invasion [39]. The uNK cells are the major source of proteases in the uterine decidua [40], mainly on the production of MMP-2 [40], which is demonstrably involved in vascular smooth muscle cells migration, important process for vascular remodeling [41, 42]. Indeed, MMP-2 and MMP-9 plays a role in the uteroplacental and vascular remodeling during healthy pregnancy and the uNK cells number reduce after week 20 of pregnancy when the spiral artery remodeling is complete [43]. In addition to the influence of uNK cells on spiral artery remodeling, trophoblast cells and vascular smooth muscle cells can produce MMP-12 that could mediate proteolysis [44].

MMP-12 is an elastolytic protease expressed by trophoblast, which is capable of degrading several extracellular matrix components, including collagen type IV, laminin, fibronectin, vitronectin and heparin sulphate proteoglycans. Furthermore, in other examples of vessel remodeling MMP-12 is capable of activating pro-MMP-2 and pro-MMP-3, therefore the role of extravillous trophoblast cells is crucial for vascular remodeling [45].

The decidualization phenomenon that is characterized by cellular differentiation of elongated spindleshaped stromal fibroblast [5] that acquires an rounded epithelial-like phenotype, instead of completing transition into an epithelial cell phenotype, since they do not express molecules of the keratin family (major epithelial cell lineage makers). Before the decidualization complete, a reduction in the extracellular matrix content occurs [46], in a time matched of MMP-2 and MMP-9 expression [47], being, therefore, the gelatinases most produced by the endometrial stromal cells in that period. It is noteworthy that before the decidualization occurs, the expression of MMP-3 can be observed, but its expression is diminished with the initiation of decidualization [48].

During the decidualization, it is not only the morphology of the stromal cells that changes, but also its task, as they acquire new essential functions to coordinate the trophoblastic invasion for the placental formation.

TROPHOBLAST INVASION

After the implantation process, the receptive endometrium undergoes an extensive invasion by the trophoblast cells throughout the early pregnancy period, until reaching the access of maternal spiral arteries, and replace the endothelium of the arterial wall forming endovascular-trophoblast [49, 50]. For trophoblast invasion to be possible, remodeling of the extracellular matrix is essential, configuring this as active biochemical process.

Stromal cells seem to express TIMP-1-3 to regulate trophoblast invasion [51]. On the other hand, trophoblast cells express MMPs for endometrial invasion, while plasmin and immune factors play important roles in the structural reorganization of the stroma. *In vitro* studies suggest the critical involvement of gelatinases in trophoblast invasion [52, 53]. The MMP-2 is the main gelatinase expressed by trophoblast cells during the early first trimester, when these cells exhibit an excessively invasive behavior. After that, MMP-9 prevails in the late first trimester of human pregnancy [54, 55]. Though MMP-2 and MMP-9 have significant effect on the regulation of trophoblast invasion in humans but, knockout mice, deficient in MMP-2 or MMP-9 are fertile and only mild effects have been reported [56, 57].

Besides, studies have shown that resistin increases migration and invasiveness of BeWo cells via upregulation of MMP-2 activity and expression, as well as suppression of tissue inhibitors of metalloproteinases 1 and 2 (TIMP-1/2) [57]. Resistin improves endothelial cell tube formation as well as expression of VEGF, the VEGF receptors, MMP-1 and MMP-2, and thus may play a role in villous angiogenesis [58, 59].

Endometrial cells can produce a hormone, the adiponectin, which have a defined role increasing the invasiveness of first trimester extravillous trophoblast cells through increase in the expression of MMP-2 and MMP-9 and decrease the expression of TIMP-2 [60]. The production of these two

MMPs can be downregulated by progesterone [61], but can be upregulated (including the MMP-3), *in vitro*, by addition of folate. Interestingly, folic acid deficiency in early pregnancy is associated with early gestational loss [61].

MMP-2 and MMP-9 production (and the cell invasiveness) by extravillous trophoblast can be inhibited by several mechanisms involving uNK cells, including TGF-β1, TNF-α and IFN-y production [62]. These two MMPs are the most studied during the trophoblast invasive procedure, however MMP-12, MMP-14 and MMP-15, as well as TIMP-1 and -2 inhibitors are also expressed during this period and may have relevant participation [63-65].

LATE PREGNANCY AND PARTURITION

At the end of gestation, an early event in the parturition takes place (several weeks before the onset of active labor), with an important structural and cervical extracellular matrix composition change, known as cervical ripening, where loss of structural integrity associated with loss of tensile strength occurs [66, 67].

Changes in the structural organization of extracellular matrix, including collagen reduction, which are mediated by an increase in hydrophilic glycosaminoglycans, especially decorin by binding to collagen fibers and disturbs the fibrillary organization, makes the uterine cervix soft, thin and easily distensible, and its load capacity decreases [68, 69]. In addition to decorin, hyaluronate also acts to change the collagen architecture by weakening its interaction with fibronectin, generating a water attraction in the spaces created, resulting in hydration and collagen dispersion [66, 70-73].

The aforementioned events are prevented by the action of progesterone, which also increases the expression of TIMP-2, reducing the action of MMPs by endocervical fibroblasts [74-77], besides antagonizing the expression of collagenases from these cells [78, 79]. The progesterone withdrawal concomitant with the increase of estrogen and prostaglandin change this scenario towards the onset of labor [80], via increased collagenase expression in endocervical cells.

The collagenases (MMP-1, MMP-8 and MMP-13) are important MMPs for the process of cervical dilation that triggers parturition. Dilation also involves a massive influx of leukocytes into the cervical stroma, and these inflammatory cells are a major source of collagenase [81].

At the time of active labor, cervical resistance to tensile force collapses, a positive feedback system is established, so that cervical distension stimulates the synthesis and secretion of oxytocin (by the posterior hypothalamic-pituitary system), which in turn acts on its myometrial receptors, increasing contraction and distension. Concomitantly, levels of inflammatory factors and prostaglandins increase in uterine tissue as a result of contractions. These factors induce the functional activation of estrogen by up-regulating the oxytocin receptors in the myometrium, making this muscle tissue more and more contractile. This positive feedback system ends with childbirth.

Among the mechanical components critical to childbirth, we have the decidual activation, which includes the onset of uterine contractions mentioned above, and rupture of fetal membranes. The biochemical mechanisms that contribute to membrane rupture include both matrix remodeling and cellular apoptosis [82].

Since collagen is the main extracellular component that maintain tensile strength (type I on mesoderm layer and type IV on basement membrane and between amnion epithelium and mesenchyme) [83, 84], the molecules directed to the degradation of this component are of extreme importance for the membrane rupture [83].

The major isoform prior to onset of labor is the MMP-1 [85], while MMP-2 is present but does not show changes in its expression in membrane rupture or parturition processes [84, 86]. Different from that observed with respect to MMP-9, which has increased expression in response to prostaglandins and Tumor Necrosis Factor [87, 88], representing, therefore, a biomarker for the imminence of membrane rupture [89, 90]. In fact, the main regulator of MMP-9 is TIMP-1, which has its reduced expression at birth, or in preterm labor [84].

The generation of inflammatory mediators as a result of an intrauterine infection leads to an early production of MMPs, which accelerates the degradation of extracellular matrix and the cervix ripening [91, 92], drastically increasing the chances of preterm birth. A number of studies have shown that preterm labor is associated with elevated MMP-9 and MMP-3 in amniotic fluid and fetal membranes [93, 94]. Moreover, placental MMP-2 and MMP-3 levels are reportedly higher in women delivering preterm compared with term [95].

The literature demonstrates the importance of the orchestrated regulation of MMPs and TIMPs in the chronology of gestation, and their imbalance or temporal failure are associated with obstetric complications. Hypertensive disorders, such as chronic hypertension, gestational hypertension and preeclampsia, are among the most common gestational complications, with indications pointing to an origin related to abnormal trophoblastic invasiveness [96].

In addition, low concentration of pro-angiogenic factors (such as VEGF and PGF), or high concentration of anti-angiogenic factors (such as sFlt-1) are involved in the process of establishing hypertensive conditions during pregnancy. It is interesting to note that sFlt-1 attenuates the activity (and concentration) of MMP-2 and MMP-9 in uterine, placental and vascular tissues, while VEGF reversed the sFlt-1 induced decreases in MMPs amount/activity [97-105]. Other studies indicate reduced expression of MMP-2, MMP-8, MMP-9 and MMP-11 in preeclampsia placentas and from Fetal Growth Restriction placental samples [106].

Studies in animal models on recurrent pregnancy loss indicate reduced expression or activity of MMP-2 and MMP-9 in the uterus, associated with increased collagen fibers deposition, which can lead to impairment of smooth muscle cell growth and cell proliferation and migration, as well as reduce uteroplacental invasiveness and reduce placental growth [107].

Considering that the orchestrated expression of MMPs and TIMPs are crucial to the normal physiology of gestation, and that changes in the expressions of these components are associated with gestational complications. Studies on these remodeling molecules of the extracellular matrix are crucial for both understanding the gestational complications and for the maternal-fetal prognosis.

It is known that any flaws in the gestational processes mentioned in this chapter result in immediate consequences for developing offspring as well as in health complications for the mother. In addition, recent studies demonstrate that these failures can have late consequences also for the mother and especially for the offspring in adult life. Targeting MMPs could provide a new approach for boost the development of strategies to prevent complications in woman's health, obstetrics and for consequences in offspring adulthood.

References

[1] Ramirez, F. & Rifkin, D. B. (2003). "Cell signaling events: a view from the matrix." *Matrix Biol*, *22*(2), 101–7.

[2] Berrier, A. L. & Yamada, K. M. (2007). "Cell-matrix adhesion." *J Cell Physiol.*, *213*(3), 565–73.

[3] Hynes, R. O. (2009). "The extracellular matrix: not just pretty fibrils." *Science.*, *326*(5957), 1216–9.

[4] Visse, R. & Nagase, H. (2003). "Matrix metalloproteinases and tissue inhibitors of metalloproteinases: structure, function, and biochemistry." *Circ Res.*, *92*(8), 827-839.

[5] Abrahamsohn, P. A. & Zorn, T. M. (1993). "Implantation and decidualization in rodents." *J Exp Zool.*, *266*(6), 603-28.

[6] Wilcox, A. J., Baird, D. D. & Weinberg, C. R. (1999). "Time of implantation of the conceptus and loss of pregnancy." *N. Engl. J. Med.*, *340*, 1796-9.

[7] Zavan, B., Amarante-Paffaro, A. M. & Paffaro, Jr. V. A. (2013). "Role of laminin on uterine remodeling and embryo development during pregnancy." In *Laminins: Structure, Biological Activity and Role in Disease*, edited by Adams DC & Garcia EO, 117-124. Nova Science Publishers, Inc.

[8] Psychoyos, A. (1986). "Uterine receptivity for nidation." *Ann. N.Y. Acad. Sci.*, *476*, 36-42.

[9] Yagel, S., Parhar, R. S., Jeffrey, J. J. & Lala, P. K. (1988). "Normal nonmetastatic human trophoblast cells share *in vitro* invasive properties of malignant cells." *J Cell Physiol.*, *136*(3), 455-62.

[10] Aplin, J. D., Charlton, A. K. & Ayad, S. (1988). "An immunohistochemical study of human endometrial extracellular matrix during the menstrual cycle and first trimester of pregnancy." *Cell Tissue Res*, *253*, 231–40.

[11] Cross, J. C., Werb, Z. & Fisher, S. J. (1994). "Implantation and the placenta: key pieces of the development puzzle." *Science*, *266*, 1508–18.

[12] Vu, T. H. & Werb, Z. (2000). "Matrix metalloproteinases: effectors of development and normal physiology." *Genes Dev*, *14*, 2123–33.

[13] Das, S. K., Yano, S., Wang, J., Edwards, D. R., Nagase, H. & Dey, S. K. (1997). "Expression of matrix metalloproteinases and tissue inhibitors of metalloproteinases in the mouse uterus during the peri-implantation period." *Dev Genet*, *21*, 44–54.

[14] Tanaka, Y., Park, J. H., Tanwar, P. S., et al. (2012). "Deletion of tuberous sclerosis 1 in somatic cells of the murine reproductive tract causes female infertility." *Endocrinology*, *153*, 404–16.

[15] Guertin, D. A. & Sabatini, D. M. (2007). "Defining the role of mTOR in cancer." *Cancer Cell*, *12*, 9–22.

[16] Hulboy, D. L., Rudolph, L. A. & Matrisian, L. M. (1997). "Matrix metalloproteinases as mediators of reproductive function." *Mol. Hum. Reprod.*, *3*, 27–45.

[17] Huppertz, B., Kertschanska, S., Demir, A. Y., Frank, H. G. & Kaufmann, P. (1998). "Immunohistochemistry of matrix metalloproteinases (MMP), their substrates, and their inhibitors (TIMP) during trophoblast invasion in the human placenta." *Cell Tissue Res.*, *291*, 133–148.

[18] Jokimaa, V., Oksjoki, S., Kujari, H., Vuorio, E. & Anttila, L. (2002). "Altered expression of genes involved in the production and degradation of endometrial extracellular matrix in patients with

unexplained infertility and recurrent miscarriages." *Mol Hum Reprod*, *8*, 1111–6.

[19] Fernandez-Patron, C., Radomski, M. W. & Davidge, S. T. (1999). "Vascular matrix metalloproteinase-2 cleaves big endothelin-1 yielding a noval vasoconstrictor." *Circ Res*, *85*, 906-11.

[20] Martin, D. C., Fowlkes, J. L., Babic, B. & Khokha, R. (1999). "Insulin-like growth factor II signaling in neoplastic proliferation is blocked by transgenic expression of the metalloproteinase inhibitor TIMP-1." *J Cell Biol*, *146*, 881-92.

[21] Yu, Q. & Stamenkovic, I. (2004). "Cell surface-localized matrix metalloproteinases-9 proteolytically activates THF-beta and promotes tumor invasion and angiogenesis." *Genes Dev*, *14*, 163-76.

[22] Bulmer, J. N., Morrison, L., Longfellow, M., Ritson, A. & Pace, D. (1991). "Granulated lymphocytes in human endometrium: histochemical and immunohistochemical studies." *Hum. Reprod.*, *6*, 791–798.

[23] Choudhury, R. H., Dunk, C. E., Lye, S. J., Aplin, J. D., Harris, L. K. & Jones, R. L. (2017). "Extravillous Trophoblast and Endothelial Cell Crosstalk Mediates Leukocyte Infiltration to the Early Remodeling Decidual Spiral Arteriole Wall." *J Immunol.*, *198*(10), 4115-4128.

[24] Moffett, A. & Colucci, F. (2015). "Co-evolution of NK receptors and HLA ligands in humans is driven by reproduction." *Immunol. Rev.*, *267*, 283–297.

[25] King, A. & Loke, Y. W. (1991). "On the nature and function of human uterine granular lymphocytes." *Immunology Today.*, *12*, 432–435.

[26] Erlebacher, A. (2013). "Immunology of the maternal–fetal interface." *Annual Review of Immunology.*, *31*, 387–411.

[27] Wallace, A. E., Fraser, R. & Cartwright, J. E. (2012). "Extravillous trophoblast and decidual natural killer cells: a remodelling partnership." *Hum. Reprod.* Update, *18*, 458–471.

[28] Langer, N., Beach, D. & Lindenbaum, E. S. (1999). "Novel hyperactive mitogen to endothelial cells: human decidual NKG5." *American Journal of Reproductive Immunology.*, *42*, 263–272.

[29] Li, X. F., Charnock-Jones, D. S., Zhang, E., Hiby, S., Malik, S., Day, K., Licence, D., Bowen, J. M., Gardner, L., King, A., et al. (2001). "Angiogenic growth factor messenger ribonucleic acids in uterine natural killer cells." *Journal of Clinical Endocrinology and Metabolism.*, *86*, 1823–1834.

[30] Hanna, J., Goldman-Wohl, D., Hamani, Y., Avraham, I., Greenfield, C., Natanson-Yaron, S., et al. (2006). "Decidual NK cells regulate key developmental processes at the human fetal-maternal interface." *Nat Med.*, *12*(9), 1065-74.

[31] Lash, G. E., Naruse, K., Robson, A., Innes, B. A., Searle, R. F., Robson, S. C. & Bulmer, J. N. (2011). "Interaction between uterine natural killer cells and extravillous trophoblast cells: effect on cytokine and angiogenic growth factor production." *Human Reproduction.*, *26*, 2289–2295.

[32] Chen, H. L., Yelavarthi, K. K. & Hunt, J. S. (1993). "Identification of transforming growth factor-beta 1 mRNA in virgin and pregnant rat uteri by *in situ* hybridization." *Journal of Reproductive Immunology.*, *25*, 221–233.

[33] Hunt, J. S. (1994). "Immunologically relevant cells in the uterus." *Biology of Reproduction.*, *50*, 461–466.

[34] Burnett, T. G. & Hunt, J. S. (2000). "Nitric oxide synthase-2 and expression of perforin in uterine NK cells." *Journal of Immunology.*, *164*, 5245–5250.

[35] Wang, C., Umesaki, N., Nakamura, H., Tanaka, T., Nakatani, K., Sakaguchi, I., Ogita, S. & Kaneda, K. (2000). "Expression of vascular endothelial growth factor by granulated metrial gland cells in pregnant murine uteri." *Cell and Tissue Research*, *300*, 285–293.

[36] Wang, C., Tanaka, T., Nakamura, H., Umesaki, N., Hirai, K., Ishiko, O., Ogita, S. & Kaneda, K. (2003). "Granulated metrial gland cells in the murine uterus: localization, kinetics, and the functional role in angiogenesis during pregnancy." *Microscopy Research and Technique.*, *60*, 420–429.

[37] Tayade, C., Hilchie, D., He, H., Fang, Y., Moons, L., Carmeliet, P., Foster, R. A. & Croy, B. A. (2007). "Genetic deletion of placenta

growth factor in mice alters uterine NK cells." *J Immunol.*, *178*, 4267-4275.

[38] Degaki, K. Y., Chen, Z., Yamada, A. T. & Croy, B. A. (2012). "Delta-like ligand (DLL) 1 expression in early mouse decidua and its localization to uterine natural killer cells." *PLoS ONE.*, *7*, 52037.

[39] Lash, G. E., Naruse, K., Innes, B. A., Robson, S. C., Searle, R. F. & Bulmer, J. N. (2010). "Secretion of angiogenic growth factors by villous cytotrophoblast and extravillous trophoblast in early human pregnancy." *Placenta*, *31*, 545-8.

[40] Naruse, K., Lash, G. E., Bulmer, J. N., Innes, B. A., Otun, H. A., Searle, R. F. & Robson, S. C. (2009). "The urokinase plasminogen activator (uPA) system in uterine natural killer cells in the placental bed during early pregnancy." *Placenta*, *30*, 398-404.

[41] Belo, V. A., Guimaraes, D. A. & Castro, M. M. (2016). "Matrix metalloproteinase 2 as a potential mediator of vascular smooth muscle cell migration and chronic vascular remodeling in hypertension." *J. Vasc. Res.*, *52* (4), 221e231.

[42] Song, H., Cheng, Y., Bi, G., Zhu, Y., Jun, W., Ma, W., et al., (2016). "Release of matrix Metalloproteinases-2 and 9 by S-Nitrosylated Caveolin-1 contributes to degradation of extracellular matrix in tpa-treated hypoxic endothelial cells." *PLoS ONE*, *11*(2), e0149269.

[43] Bulmer, J. N. & Lash, G. E. (2005). "Human uterine natural killer cells: a reappraisal." *Mol Immunol*, *42*, 511-21.

[44] Harris, L. K., Smith, S. D., Keogh, R. J., Jones, R. L., Baker, P. N., Knofler, M., et al. (2010). "Trophoblast- and vascular smooth muscle cell-derived MMP-12 mediates elastolysis during uterine spiral artery remodeling." *Am J Pathol.*, *177*, 2103–15.

[45] Lagente, V., Le Quement, C. & Boichot, E. (2009). "Macrophage metalloelastase (MMP-12) as a target for inflammatory respiratory diseases." *Expert Opin Ther Targets*, *13*, 287–295.

[46] Rider, V., Carlone, D. L., Witrock, D., Cai, C. & Oliver, N. (1992). "Uterine fibronectin mRNA content and localization are modulated during implantation." *Dev. Dyn.*, *195*, 1-14.

[47] Bany, B. M., Harvey, M. B. & Schultz, G. A. (2000). "Expression of matrix metalloproteinases 2 and 9 in the mouse uterus during implantation and oil-induced decidualization." *J Reprod Fertil*, *120*(1), 125–34.

[48] Duval, F., Dos Santos, E., Moindjie, H., Serazin, V., Swierkowski-Blanchard, N., Vialard, F. & Dieudonné, M. N. (2017). "Adiponectin limits differentiation and trophoblast invasion in human endometrial cells." *J Mol Endocrinol.*, *59*(3), 285-297.

[49] Loke, Y. W. & King, A. (1995). "Human trophoblast development." In: *Human implantation: cell biology and immunology.* Cambridge: Cambridge University Press, 1995, p. 32-62.

[50] Kaufmann, P. & Scheffen, I. (1992). "Placental development." In: *Fetal and neonatal physiology.* Edited by: Polin RA, Fox WW. Philadelphia: W.B. Saunders. p. 47-56.

[51] Popovici, R. M., Betzler, N. K., Krause, M. S., et al. (2006). "Gene expression profiling of human endometrial-trophoblast interaction in a coculture model." *Endocrinology*, *147*, 5662–75.

[52] Librach, C. L., Werb, Z., Fitzgerald, M. L., Chiu, K., Corwin, N. M., Esteves, R. A., Grobelny, D., Galardy, R., Damsky, C. H. & Fisher, S. J. (1991). "92-kD type IV collagenase mediates invasion of human cytotrophoblasts." *J Cell Biol*, *113*, 437-49.

[53] Isaka, K., Usuda, S., Ito, H., Sagawa, Y., Nakamura, H., Nishi, H., Suzuki, Y., Li, Y. F. & Takayama, M. (2003). "Expression and activity of matrix metalloproteinase 2 and 9 in human trophoblasts." *Placenta*, *24*, 53-64.

[54] Xu, P., Wang, Y., Zhu, S., Luo, S., Piao, Y. & Zhuang, L. (2000). "Expression of matrix metalloproteinase-2, -9 and -14, tissue inhibitors of metalloproteinse-1, and matrix proteins in human placenta during the first trimester." *Biol Reprod*, *62*, 988-94.

[55] Staun-Ram, E., Goldman, S., Gabarin, D. & Shalev, E. (2004). "Expression and importance of matrix metalloproteinases 2 and 9 (MMP-2 and -9) in human trophoblast invasion." *Reprod Biol Endocrinol*, *2*, 59.

[56] Itoh, T., Ikeda, T., Gomi, H., Nakao, S., Suzuki, T. & Itohara, S. (1997). "Unaltered secretion of beta amaloid precursor protein in gelatinase A (matrix metalloproteinase 2)-deficient mice." *J Biol Chem*, *272*, 22389-92.

[57] Dubois, B., Arnold, B. & Opdenakker, G. (2000). "Gelatinase B deficiency impairs reproduction." *J Clin Invest*, *106*, 627-8.

[58] Di Simone, N., Di Nicuolo, F., Sanguinetti, M., et al. (2006). "Resistin regulates human choriocarcinoma cell invasive behaviour and endothelial cell angiogenic processes." *J Endocrinol*, *189*, 691–9.

[59] D'Ippolito, S., Tersigni, C., Scambia, G. & Di, S. N. (2012). "Adipokines, an adipose tissue and placental product with biological functions during pregnancy." *Biofactors*, *38*, 14–23.

[60] Benaitreau, D., Dos Santos, E., Leneveu, M. C., Alfaidy, N., Feige, J. J., de Mazancourt, P., Pecquery, R. & Dieudonné, M. N. (2010). "Effects of adiponectin on human trophoblast invasion." *J Endocrinol*, *207*, 45-53.

[61] Chen, J. Z., Wong, M. H., Brennecke, S. P. & Keogh, R. J. (2011). "The effects of human chorionic gonadotrophin, progesterone and oestradiol on trophoblast function." *Mol Cell Endocrinol*, *342*, 73-80.

[62] Lash, G. E., Schiessl, B., Kirkley, M., Innes, B. A., Cooper, A., Searle, R. F., Robson, S. C. & Bulmer, J. N. (2006). "Expression of angiogenic growth factors by uterine natural killer cells during early pregnancy." *J Leukoc boil*, *80*, 572-80.

[63] Anacker, J., Segerer, S. E., Hagemann, C., Feix, S., Kapp, M., Bausch, R. & Kammerer, U. (2011). "Human decidua and invasive trophoblasts are rich sources of nearly all human matrix metalloproteinases." *Molecular Human Reproduction*, *17*(10), 637–652.

[64] Bischof, P., Meisser, a. & Campana, a. (2002). "Control of MMP-9 expression at the maternal-fetal interface." *Journal of Reproductive Immunology*, *55*(1–2), 3–10.

[65] Cohen, M., Meisser, A. & Bischof, P. (2006). "Metalloproteinases and human placental invasiveness." *Placenta*, *27*(8), 783–93.

[66] Leppert, P. C. (1995). "Anatomy and physiology of cervical ripening." *Clin Obstet Gynecol*, *38*, 267–79.

[67] Word, R. A., Li, X. H., Hnat, M. & Carrick, K. (2007). "Dynamics of cervical remodeling during pregnancy and parturition: mechanisms and current concepts." *Semin Reprod Med*, *25*, 69–79.

[68] Kokenyesi, R. & Woessner, Jr. J. F. (1991). "Effects of hormonal perturbations on the small dermatan sulfate proteoglycan and mechanical properties of the uterine cervix of late pregnant rats." *Connect Tissue Res*, *26*, 199–205.

[69] Leppert, P. C., Kokenyesi, R., Klemenich, C. A. & Fisher, J. (2000). "Further evidence of a decorin–collagen interaction in the disruption of cervical collagen fibers during rat gestation." *Am J Obstet Gynecol*, *182*, 805–11.

[70] Ludmir, J. & Sehdev, H. M. (2000). "Anatomy and physiology of the uterine cervix." *Clin Obstet Gynecol*, *43*, 433–9.

[71] Straach, K. J., Shelton, J. M., Richardson, J. A., Hascall, V. C. & Mahendroo, M. S. (2005). "Regulation of hyaluronan expression during cervical ripening." *Glycobiology*, *15*, 55–65.

[72] Soh, Y. M., Tiwari, A., Mahendroo, M., Conrad, K. P. & Parry, L. J. (2012). "Relaxin regulates hyaluronan synthesis and aquaporins in the cervix of late pregnant mice." *Endocrinology*, *153*, 6054–64.

[73] Ruscheinsky, M., De la Motte, C. & Mahendroo, M. (2008). "Hyaluronan and its binding proteins during cervical ripening and parturition: dynamic changes in size, distribution and temporal sequence." *Matrix Biol*, *27*, 487–97.

[74] Imada, K., Ito, A., Sato, T., Namiki, M., Nagase, H. & Mori, Y. (1997). "Hormonal regulation of matrix metalloproteinase 9/gelatinase B gene expression in rabbit uterine cervical fibroblasts." *Biol Reprod*, *56*, 575–80.

[75] Sato, T., Ito, A., Mori, Y., Yamashita, K., Hayakawa, T. & Nagase, H. (1991). "Hormonal regulation of collagenolysis in uterine cervical fibroblasts. Modulation of synthesis of procollagenase, prostromelysin and tissue inhibitor of metalloproteinases (TIMP) by progesterone and oestradiol-17 beta." *Biochem J*, *275*, 645–50.

[76] Ito, A., Imada, K., Sato, T., Kubo, T., Matsushima, K. & Mori, Y. (1994). "Suppression of interleukin 8 production by progesterone in rabbit uterine cervix." *Biochem J*, *301*, 183–6.

[77] Imada, K., Ito, A., Itoh, Y., Nagase, H. & Mori, Y. (1994). "Progesterone increases the production of tissue inhibitor of metalloproteinases-2 in rabbit uterine cervical fibroblasts." *FEBS Lett*, *341*, 109–12.

[78] Rajabi, M. R., Dodge, G. R., Solomon, S. & Poole, A. R. (1991). "Immunochemical and immunohistochemical evidence of estrogen-mediated collagenolysis as a mechanism of cervical dilatation in the guinea pig at parturition." *Endocrinology*, *128*, 371–8.

[79] Rajabi, M., Solomon, S. & Poole, A. R. (1991). "Hormonal regulation of interstitial collagenase in the uterine cervix of the pregnant guinea pig." *Endocrinology*, *128*, 863–71.

[80] Osmers, R., Rath, W., Pflanz, M. A., Kuhn, W., Stuhlsatz, H. W. & Szeverenyi, M. (1993). "Glycosaminoglycans in cervical connective tissue during pregnancy and parturition." *Obstet Gynecol*, *81*, 88–92.

[81] Osmers, R., Rath, W., Adelmann-Grill, B. C., et al. (1992). "Origin of cervical collagenase during parturition." *Am J Obstet Gynecol*, *166*, 1455–60.

[82] Moore, R. M., Mansour, J. M., Redline, R. W., Mercer, B. M. & Moore, J. J. (2006). "The physiology of fetal membrane rupture: insight gained from the determination of physical properties." *Placenta*, *27*, 1037–51.

[83] Menon, R. & Fortunato, S. J. (2004). "The role of matrix degrading enzymes and apoptosis in rupture of membranes." *J Soc Gynecol Invest*, *11*, 427–37.

[84] Bachmaier, N. & Graf, R. (1999). "The anchoring zone in the human placental amnion: bunches of oxytalan and collagen connect mesoderm and epithelium." *Anat Embryol (Berl).*, *200*(1), 81-90.

[85] Bryant-Greenwood, G. D. & Yamamoto, S. Y. (1995). "Control of peripartal collagenolysis in the human chorion-decidua." *Am J Obstet Gynecol*, *172*, 63–70.

[86] Xu, P., Alfaidy, N. & Challis, J. R. (2002). "Expression of matrix metalloproteinase (MMP)-2 and MMP-9 in human placenta and fetal membranes in relation to preterm and term labor." *J Clin Endocrinol Metab*, *87*, 1353–61.

[87] McLaren, J., Taylor, D. J. & Bell, S. C. (2000). "Prostaglandin E(2)-dependent production of latent matrix metalloproteinase-9 in cultures of human fetal membranes." *Mol Hum Reprod*, *6*, 1033–40.

[88] Ulug, U., Goldman, S., Ben-Shlomo, I. & Shalev, E. (2001). "Matrix metalloproteinase (MMP)-2 and MMP-9 and their inhibitor, TIMP-1, in human term decidua and fetal membranes: the effect of prostaglandin F(2alpha) and indomethacin." *Mol Hum Reprod*, *7*, 1187–93.

[89] El Khwad, M., Pandey, V., Stetzer, B., et al. (2006). "Fetal membranes from term vaginal deliveries have a zone of weakness exhibiting characteristics of apoptosis and remodeling." *J Soc Gynecol Invest*, *13*, 191–5.

[90] Kumar, D., Fung, W., Moore, R. M., et al. (2006). "Proinflammatory cytokines found in amniotic fluid induce collagen remodeling, apoptosis, and biophysical weakening of cultured human fetal membranes." *Biol Reprod*, *74*, 29–34.

[91] Cockle, J. V., Gopichandran, N., Walker, J. J., Levene, M. I. & Orsi, N. M. (2007). "Matrix metalloproteinases and their tissue inhibitors in preterm perinatal complications." *Reprod Sci.*, *14*(7), 629-645.

[92] Goldenberg, R. L., Hauth, J. C. & Andrews, W. W. (2000). "Intrauterine infection and preterm delivery." *N Engl J Med.*, *342*(20), 1500-1507.

[93] Maymon, E., Romero, R., Pacora, P., Gervasi, M. T. & Gomez, R. (2000). "Evidence of *in vivo* differential bioavailability of the active forms of matrix metalloproteinases 9 and 2 in parturition, spontaneous rupture of membranes, and intraamniotic infection." *Am J Obstet Gynecol.*, *183*(4), 887-894.

[94] Fortunato, S. J., Menon, R. & Lombardi, S. J. (1999). "Stromelysins in placental membranes and amniotic fluid with premature rupture of membranes." *Obstet Gynecol.*, *94*(3), 435-440.

[95] Sundrani, D., Chavan-Gautam, P., Pisal, H., Mehendale, S. & Joshi, S. (2013). "Matrix metalloproteinases-2, -3 and tissue inhibitors of metalloproteinases-1, -2 in placentas from preterm pregnancies and their association with one-carbon metabolites." *Reproduction.*, *145*(4), 401-410.

[96] Roberts, J. M. & Escudero, C. (2012). "The placenta in preeclampsia." *Pregnancy Hypertens.*, *2*, 72–83.

[97] Karumanchi, S. A. & Bdolah, Y. (2004). "Hypoxia and sFlt-1 in preeclampsia: the "chicken-and-egg" question." *Endocrinology.*, *145*, 4835–7.

[98] Wolf, M., Shah, A., Lam, C., Martinez, A., Smirnakis, K. V., Epstein, F. H., et al. (2005). "Circulating levels of the antiangiogenic marker sFLT-1 are increased in first versus second pregnancies." *Am J Obstet Gynecol.*, *193*, 16–22.

[99] Rajakumar, A., Michael, H. M., Rajakumar, P. A., Shibata, E., Hubel, C. A., Karumanchi, S. A., et al. (2005). "Extraplacental expression of vascular endothelial growth factor receptor-1, (Flt-1) and soluble Flt-1 (sFlt-1), by peripheral blood mononuclear cells (PBMCs) in normotensive and preeclamptic pregnant women." *Placenta.*, *26*, 563–73.

[100] Rana, S., Karumanchi, S. A., Levine, R. J., Venkatesha, S., Rauh-Hain, J. A., Tamez, H., et al. (2007). "Sequential changes in antiangiogenic factors in early pregnancy and risk of developing preeclampsia." *Hypertension.*, *50*, 137–42.

[101] Lam, C., Lim, K. H. & Karumanchi, S. A. (2005). "Circulating angiogenic factors in the pathogenesis and prediction of preeclampsia." *Hypertension.*, *46*, 1077–85.

[102] LaMarca, B. D., Gilbert, J. & Granger, J. P. (2008). "Recent progress toward the understanding of the pathophysiology of hypertension during preeclampsia." *Hypertension.*, *51*, 982–8.

[103] Tsatsaris, V., Goffin, F., Munaut, C., Brichant, J. F., Pignon, M. R., Noel, A., et al. (2003). "Overexpression of thesoluble vascular endothelial growth factor receptor in preeclamptic patients:

pathophysiologicalconsequences." *J Clin Endocrinol Metab.*, *88*, 5555–63.

[104] Makris, A., Thornton, C., Thompson, J., Thomson, S., Martin, R., Ogle, R., et al. (2007). "Uteroplacental ischemia results in proteinuric hypertension and elevated sFLT-1." *Kidney Int.*, *71*, 977–84.

[105] Maynard, S. E., Min, J. Y., Merchan, J., Lim, K. H., Li, J., Mondal, S., et al. (2003). "Excess placental soluble fms-like tyrosine kinase 1 (sFlt1) may contribute to endothelial dysfunction, hypertension, and proteinuria in preeclampsia." *J Clin Invest.*, *111*, 649–58.

[106] Zhu, J., Zhong, M., Pang, Z. & Yu, Y. (2014). "Dysregulated expression of matrix metalloproteinases and their inhibitors may participate in the pathogenesis of pre-eclampsia and fetal growth restriction." *Early Hum Dev.*, *90*(10), 657-64.

[107] Li, W., Mata, K. M., Mazzuca, M. Q. & Khalil, R. A. (2014). "Altered matrix metalloproteinase-2 and -9 expression/activity links placental ischemia and anti-angiogenic sFlt-1 to uteroplacental and vascular remodeling and collagen deposition in hypertensive pregnancy." *Biochem Pharmacol.*, *89*(3), 370-85.

In: A Closer Look at Metalloproteinases
Editor: Lena Goodwin
ISBN: 978-1-53616-517-3

Chapter 6

THE ROLE OF MATRIX METALLOPROTEINASES IN BONE TISSUE

Yeliz Basaran Elalmis[1,*], Cem Özel[1], Ceren Kececiler[1], Dilan Altan[1], Ecem Tiryaki[1,2], Ali Can Özarslan[1], Bilge Sema Tekerek[3] and Sevil Yücel[1]

[1]Department of Bioengineering, Yildiz Technical University, Istanbul, Turkey
[2]Department of Physical Chemistry, Universidade de Vigo, Vigo, Spain
[3]Department of Genetics and Bioengineering, Nişantaşı University, Istanbul, Turkey

ABSTRACT

Bone has a dynamic structure, since it is remodelled during the lifespan to sustain its structure and function. Extracellular matrix (ECM) is playing a tremendously important role, such as cell adhesion, immobilization of growth factors and nucleation of mineralization in bone

* Corresponding Author's E-mail: yelizbasaran@gmail.com.

development phase. It consists of proteins and leads the bone remodelling by the combined osteoblast (bone-forming cells) and osteoclast (bone-resorbing cells) activities. Besides, ECM behaves as a scaffold for mineral deposition. Matrix metalloproteinases (MMPs), a family of zinc-depended proteolytic enzymes, are the most important enzymes used for the degradation of unrelated proteins and structural components present in ECM. MMPs are highly expressed in mammalian bone and cartilage cells and are able to cleave collagens, thus function as collagenases. Furthermore, they lead remodelling of ECM in connection with tissue specific and cell anchored inhibitors. Functions of MMPs may vary bone quality via bone resorption and formation, i.e., osteoblast recruitment and survival, angiogenesis, osteocyte viability and function, chondrocyte proliferation and differentiation. Abnormal expression of MMPs can be related to pathological conditions such as unstable bone remodelling, particularly osteoporosis, rheumatoid arthritis and osteoarthritis. In this chapter, bone tissue components, MMP properties and functions, bone modelling, remodelling and resorption, repair and regeneration, and pathological bone resorption will be discussed.

Keywords: Matrix metalloproteinases (MMPs), bone remodeling, bone regeneration, Tissue inhibitors of matrix metalloproteinases (TIMPs), extracellular matrix (ECM)

1. Introduction

Bone tissue has tremendous importance since it is the main element of vertabretres skeleton. The osteoblasts involved in bone formation and the osteoclasts responsible for bone resorption continuously remodels the bone tissue throughout life. A “matrix” is required for the bone tissue cells to perform their vital activity. Extracellular matrix (ECM) is a network of macromulecules which serves structural support for tissues and provides a micro environment which cells can communicate through, via their receptors. ECM secretes molecules which are substantial for adhesion and migration of bone cells (Visse and Nagase 2003; Green et al. 2017). Degredation of the ECM components in time have a very crucial role for the development, modelling, and remodelling of the bone matrix. Matrix metalloproteinases (MMPs), are a class of zinc dependent proteases, expressed by the osteoblasts, osteocytes, osteoclasts, and chondrocytes of bone tissue and have pivotal activities during the degradation of ECM. The proper regulation of MMPs and ECM provides healty bone and cartilage formation under normal physiological conditions.

The detection of collagenases during the methamorphosis of tadpole, in 1962, has led to the idea that these enzymes may have similar activities in mammalian tissue (Gross and Lapiere 1962). The study in 1963 to elucidate the mechanism of bone resorption can be shown as a pioneer for study of MMPs on bone tissue (Woods and Nichols 1963). The MMP-deficient mouse models are most commonly used to define functions of different types of MMPs. The bone-related abnormality including osteoporosis, osteoarthritis (OA), rheumatoid arthritis (RA) etc. have been reported in case of MMPs deficiency. This can be shown as a proof that MMPs play role in processes related to precise bone formation. MMP-1, MMP-2, MMP-8, MMP-9, MMP-14, and MMP-16 have been shown to be the most active MMP variants during normal bone and cartilage development (Paiva and Granjeiro 2017). The activities of MMPs are regulated and inhibited by several endogenous inhibitors. The tissue inhibitor of matrix metalloproteinases (TIMPs), reversion-inducing-cysteine-rich protein with Kazal motifs (RECK), and α2-macroglobulin (α2M) are natural inhibitors of MMPs. The altered expression of these inhibitors has been detected during bone regeneration, resorption, and repair processes (Oh et al. 2001). The lack of balance between MMPs and their inhibitors can lead to diseases including atheroma, arthritis, RA, OA, osteoporosis, and several cancer types.

In this review, we will deal with topics related to the general aspects of MMPs and their inhibitors functions in the development, modelling and repair of bone and cartilage tisssue as well as bone-related diseases and potential usage of MMPs and inhibitors for the bioengineering applications.

2. Bone

Bone tissue is a complex structure consisting of inorganic-organic matrix components, cell-derived factors, and cells. The interactions between cells, cell-derived factors, and inorganic-organic matrix allow the bone to constantly maintain, modeling and remodeling (Buckwalter and Cooper 1987; Boskey and Posner 1984). Bone ECM comprises of minerals (such as apatite), approximately 65% by weight, and organic part (collagenous proteins and non-collagenous proteins) contributes about 25% to its composition. The rest (approximately 10%) of its composition consists of unbounded water and bounded water with collagen-mineral composite. About 90% of the organic part is composed of type I collagen with a smaller number of different collagen types and the remaining 10% is composed of non-collagenous proteins (NCPs) (Burr 2019).

Bone cells are divided into three categories: i) osteoblasts, ii) osteocytes and iii) osteoclasts. The cells, liable for bone formation (organic bone matrix secretion) and mineralization are called osteoblasts. In contrast, osteoclasts are responsible for bone resorption. The osteocytes control bone homeostasis, phosphate metabolism, and other cells functions. Osteoclasts are derived from hematopoietic stem cells (HSCs), while both osteoblasts and osteocytes are derived from mesenchymal stem cells (MSCs) (Safadi et al. 2009).

2.1. Extracellular Matrix of Bone

The bone ECM is a highly dynamic structure that helps the bone cell to communicate and it supports bone cells (Bonnans, Chou, and Werb 2014). It does not only provide mechanical support but also creates a complex and organized framework that plays an important role in bone hemostasis. In addition, bone ECM secretes various molecules that stimulate bone cells and their functions during bone formation, and resorption process (such as cell adhesion and migration) (Green et al. 2017; Deleon et al. 2014). The inorganic part of bone ECM contains overwhelmingly P and Ca ions; however, Na, K, Mg, HCO_3^-, F, Zn, and Sr are also significantly present. This matrix consists predominantly of the plate or spindle-shaped crystals of hydroxyapatite $Ca_{10}(PO_4)_6(OH)_2$ which forms due to nucleation of Ca and P ions (Bonjour 2011).

The organic part of ECM contains mainly collagenous protein type I collagen, and non-collagenous proteins include glycoproteins (thrombospondin, alkaline phosphatase, fibronectin), proteoglycans (hyaluronan, heparin sulfate), proteins of the small integrin-binding ligand N-linked glycoprotein (SIBLING), sialoproteins, osteopontin, matrix extracellular phosphoglycoprotein (MEPE), dentin matrix acidic phosphoprotein 1 (DMP-1), osteocalcin, osteonectin (Aszódi et al. 2000; Burr 2019).

Collagen is a protein composed of three alpha chains (two a_1 peptide chains and one a_2 chain) wrapped over each other. There are many studies showing that subtypes of collagen are expressed in bone. Major collagen types associated with the bone are categorized from type I to type XVII. All subtypes are expressed by different genes and are structurally different from each other. The most abundant type is collagen I (also known as an alpha-1 polypeptide) and this type is product of *Col1a1* and *Col1a2* genes (Gentili and Cancedda 2009; Velleman 2000; Safadi et al. 2009).

Cell-matrix interactions are regulated by proteolytic enzymes which are responsible for the hydrolysis of the bone ECM components. In bone, degradation of the organic ECM depends on the cysteine proteinase family or the MMP family. In addition to having essential functions during bone matrix degradation these enzymes also play a crucial role in the control of signals generated by matrix molecules by regulating the composition and integrity of the ECM structure, bone cells proliferation, differentiation, and bone cells death.

MMPs have an important place within these enzyme systems (especially bone modeling or remodeling; bone formation and bone resorption with osteoblasts and osteoclasts activity) (Reel 2006; Murphy 2015; Delaisse et al. 2003; Kusano et al. 1998; Liang et al. 2016). MMP-1, MMP-8, MMP-13, and MMP-18 are in the group of collagenases. The main characteristic of these enzymes is the ability to break down type I, II, III collagen in the intercellular space (Fields 1991; Murphy 2015). MMP-2 which is called gelatinase also cleave denatured collagen I. MMP-2 mutated people develop multicentric osteolysis (an autosomal recessive genetic disease) that results in bone destruction due to active enzyme deficiency. This situation proves that MMP-2 is important in bone formation in humans (McIntush 2004). MMP-14 can easily break down type I collagen and demonstrate pericellular collagenolytic activity. Likewise, MMP-14 contribute preosteoblast differentiation as a major collagenase *in vitro* and *Mmp14* gene expression can be arranged by ECM in order to influence osteogenic differentiation (Changlian et al. 2010).

2.2. Bone Cells

2.2.1. Osteoblasts

Osteoblasts are cuboidal cells that are known as the cells liable for bone formation. Mature osteoblasts are located on the bone surface and have morphological properties similar to cells that secrete high levels of protein. They have large nuclei located close to the basal membrane of the cell, abundant endoplasmic reticulum (ER) and a substantial amount of golgi complex as polarized cells. In addition, osteoblasts are highly differentiated cells and they have receptors to respond many hormones and local factors (Capulli, Paone, and Rucci 2014; Lester and Talmage 1981; Pockwinse et al. 2004).

Transcription factors such as Runt-related transcription factor-2 (Runx2 or known as core-binding factor α subunit/Cbfa1 or AML3), distal-less homeobox 5 (Dlx5) and Sp7/Osterix are important for progenitor cells to differentiate into osteoblasts. *Runx2* "as known master control gene" is an especially important transcription factor because of the regulation osteoblast differentiation and function by several signaling pathways. Osteoblast deficiency in *Runx2*-null mice has been shown in previous studies. Further, *Runx2* targeted deletion resulted devoid of bone formation in mice models (Rodan and Harada 1997; Ducy et al. 1997; Komori et al. 1997). *Runx2* upregulates osteoblast-related genes such as *Col1a1*, *ALP,* and *OCN* and regulate the expression of osteoblast phenotype-specific genes that control bone remodeling, comprising MMP-13 and RANK ligand; RANKL (tumour necrosis factor (TNF) ligand superfamily member II; receptor activator of the "nuclear factor kappa B, NF-kB" (RANK) ligand) (Fakhry 2013).

Osteoblastic differentiation and MSCs fate also depend on several MMP types such as MMP-14, MMP-16. While MSCs undergo osteogenic differentiation in the presence of MMP-16, the absence of MMP-14 reduces osteoblastic differentiation. Similarly, MMP-14 arranges cell morphology by some pathways and with alternation transcriptional factors, leading to affect MSCs fate (Taz et al. 2013). Osteoblasts are the primary resource of several MMP types such as MMP-1, MMP-2, MMP-3, MMP-8, MMP-9, MMP-10, MMP-11, MMP-12, MMP-13, MMP-14, and MMP-16. At most, MMP-2, MMP-13, and MMP-14 are expressed in osteoblasts.

Basically, osteoblasts exhibit three different individual phenotype stages, comprising proliferation-differentiation, bone ECM production, maturation, and mineralization. At the proliferation phase of osteoblasts, there is a high-level expression of the AP-1 family such as c-Fos, c-Jun, Jun-D, histone H4, Runx2 and Col1a1 and osteoblast progenitors which are called preosteoblasts show alkaline phosphatase (ALP) activity. Runx2 and MMP genes cooperate during osteoblastogenesis in the bone and MMP-13 is the best example of this cooperation. Previous studies have shown that *Mmp13* gene is a target gene of Runx2 in the osteoblasts as MMP-13 expression was not found in *Runx2*$^{-/-}$ mice (Jiménez et al. 1999; Paiva and Granjeiro 2017). There are several mechanisms for the role of MMP-13 in osteoblastogenesis. RUNX/RD/Cbfa complex, which is formed by the binding of the c-Fos/c-Jun to Runx2 and runt domain (RD) on activator protein (AP-1), may regulate MMP-13 expression and drives gene transcription in osteoblasts (D'Alonzo et al. 2002). Osteoblasts secrete type I collagen, non-collagenous proteins (osteocalcin (OCN), osteonectin, bone sialoprotein (BSP) I/II, osteopontin), proteoglycans and tumour growth factor-1β (TGFβ) at the ECM production phase. During the maturation, the expression of ALP, osteopontin, Cbfa1, and Osterix (Osx) increases. Thereafter, the secretion of OCN increases in the process of mineralization (Ducy and Karsenty 1999; Ducy et al. 1997).

Mineralization can be divided into steps including, the vesicular and the fibrillar phases. The matrix vesicles with a varying diameter in the 30-200 nm range are released from osteoblasts into newly formed bone regions and bind to proteoglycans in the vesicular phase. Proteoglycans immobilize Ca ions within matrix vesicles and enzymes secreted by mature osteoblasts degraded proteoglycans. Thus, Ca ions released from the proteoglycans towards the Ca channels into the matrix vesicles (Anderson 2003; Yoshiko et al. 2007; Arana-Chavez, Soares, and Katchburian 1995). On the other hand, ALP degraded P-containing compounds and P ions also pass into the matrix vesicles. Nucleation and formation of hydroxyapatite crystals occur after Ca and P migration. Then, hydroxyapatite crystals tear the matrix vesicles and spread to the surrounding matrix in the fibrillar phase. Mature osteoblasts fate may continue as osteocytes cells or bone lining cells or osteoblasts undergo apoptosis (Glimcher 1998; Boivin and Meunier 2002; Boivin et al. 2008; Parfitt 1990; Jilka et al. 1998).

2.2.2. Osteocytes

Some mature osteoblasts have undergone morphological changes, lose lots of organelles (such as golgi apparatus and rough endoplasmic reticulum) and turn into star-shaped cells during bone formation in order to coordinate the function of all bone cells and ECM factors. In this way, osteocytes represent the differentiated and morphologically altered mature osteoblasts (amount; up to 20%) that have embedded throughout the mineralized bone ECM. Osteocytes are the most abundant in bone ECM and on the bone surface. They communicate osteoblasts, lining cells, osteoclasts, and other osteocytes and establish communication network (osteocytic network) between cells. This cell network ensures the detection of local bone damages and determines the demands of bone formation or degradation (Weinger and Holtrop 1974; Aubin and Liu 1996; Bonewald 2007).

Osteocytes located particular void, called lacunae, where the cells connect and exhibit dendritic processes through narrow canaliculi and gap junctions inside the bone mineralized matrix (Parfitt 1977). Gap junctions are appropriate points that provide neighboring osteocytes or other bone cells to contact and enable the intercellular transport of various signal molecules. Similarly, proteins produced by osteocytes and secreted into the vascular system are readily delivered to cells by communication systems, including lakuna-canaliculin (Dallas, Prideaux, and Bonewald 2013; Civitelli et al. 2007).

Osteoblasts and osteoclasts are temporary cells located on different parts of bone in low numbers and they show activity during defect repairs and remodeling process. However, osteocytes are permanent bone cells and found in every part of the bone. In this way, osteocytes may be named the mechanosensory cells as of having specific communication systems and strategic locations, which allow the variation detection of mechanical signals (such as fluid flow) and number of factors in circulation (such as hormones or ions). Thus, they stimulate osteoblasts and osteoclasts functions to modify microenvironment and regulate cell activities and mineral homeostasis for adaptation of bone against environmental changes (such as mechanical forces, shear stresses, etc.). Additionally, there are several molecules for mechanical stimulation of bone cells such as adenosine triphosphate (ATP), nitric oxide (NO), calcium (Ca) and prostaglandins (PGE_1 and PGE_2) (Rochefort, Pallu, and Benhamou 2010; Burger and Klein-Nulend 1999; Weinbaum, Cowin, and Zeng 1994).

Osteocytic gene expression determines morphological and functional changes of bone cells and bone matrix. During the osteoblast to osteocyte differentiation, E11/PDPN/GP38, CD44 and plastin/fimbrin which are related to osteocyte development and maturation plays an important role. These proteins are highly expressed in osteocytes for dendritic formation and branching of osteocytes (Zhang et al. 2006). The bone matrix degradation is expressed in the canalicular extension, MMP-14 facilitates the embedding of cells into bone matrix, MMP-14 is also directly associated with osteocyte differentiation, which is one of the fate of osteoblasts. Also, MMP-14 is important for osteoblast/osteocyte transition and matrix mineralization because of controlling TGF-β activation and controlling MMPs upregulation or downregulation. MMP-11, MMP-14, and MMP-19 appear to be upregulated in bone mineralization. However, MMP-2, MMP-28, MMP-23 appear to be downregulated in osteocytes differentiation (Karsdal et al. 2004; Prideaux et al. 2015). On the other hand, when osteoblast-osteocytes differentiation was accomplished, and mature osteocytes located in the mineralized bone matrix, ALP and type I collagen expressed lower, whereas OCN expression is higher in osteocytes. In the phosphate metabolism and matrix mineralization, phosphate-regulating neutral endopeptidase (PEX), matrix extracellular phosphoglycoprotein (MEPE) are expressed by osteocytes to regulate phosphate homeostasis and to inhibit bone formation, respectively. Furthermore, dentin matrix acidic phosphoprotein 1 (DMP-1) and fibroblast growth factor 23 (FGF-23) which are expressed by early and mature osteocytes, have a crucial function in phosphate metabolism and mineralization. In particularly, DMP-1 downregulates the expression of FGF-23 for reabsorption of phosphate in order to maintain normal bone mineral content (Ubaidus et al. 2009; Yamada et al. 2004; Bonewald 2007).

Bone formation induced by systemic elevation of parathyroid hormone (PTH). In contrast, sclerostin molecules (encoded by SOST gene) are only expressed in late embedded osteocyte and these osteocyte-derived molecules inhibit the osteoblastic bone formation activity as a bone formation inhibitor (Winkler et al. 2003). RANKL, macrophage colony-stimulating factor 1 (M-CSF) and osteoprotegerin (OPG) are also expressed in osteocytes to regulate bone resorption as an activator and/or inhibitor of osteoclast differentiation (Kramer et al. 2010).

2.2.3. Osteoclasts

Osteoclasts are the giant cells that have lots of small nucleus (2 to 100 nuclei per cell) in a uniform size. The osteoclasts are responsible for the degradation of the bone during the bone modeling or remodeling, and they are differentiated from the monocyte-macrophage lineage which originated from hematopoietic cells (Boyce et al. 1999; Yavropoulou and Yovos 2008). They contain a large number of lysosomes and various lysosomal enzymes in the cytoplasm and have a large number of long-short and thick-thin cytoplasmic extensions that are directed towards the bone tissue surface. In order to resorb the inorganic and organic phases of bone ECM, the resorption pits which is called Howship's lacunae occur due to the secretion of osteoclastic protons and enzymes (such as collagenases, gelatinases etc.) from cytoplasmic extensions into the pits and then the active osteoclasts, which are ready for resorption, located in these pits (Sasaki, Debari, and Hasemi 1993; Boyle, Simonet, and Lacey 2003; Clarke 2008).

Osteoclastogenesis is directly related to bone modeling/remodeling and depends on the cooperation between all bone cells, especially the interaction between the osteoblasts and the osteoclast precursor cells are necessary. There are five phases in osteoclastogenesis; i) determination of monocytes, ii) proliferation in response to growth factors and survival of preosteoclasts (osteoclast precursor), iii) commitment and differentiation of osteoclast precursor and osteoclastic fusion (to form multinucleated cells from mononucleated cells), iv) maturation which includes polarization of osteoclasts, structural alternation, and rearrangement of active osteoclast cytoskeleton, function of osteoclasts (adhesion, migration, and resorption) in bone resorption, v) controlled-programmed cell death (apoptosis) of mature osteoclasts (Bellido, Plotkin, and Bruzzaniti 2014, 2019).

In the osteoclastogenesis, RANKL and M-CSF are fundamental factors which are expressed on the surface (as a membrane-bound form) of bone marrow stromal cells, osteoblastic cells, osteocytes and also may be available in soluble form in the microenvironment. On the other hand, activated T-lymphocyte cells express both membrane-bound and soluble forms of RANKL and secretes into microenvironment for stimulation of osteoclastogenesis (Parvizi 2010; Ross 2006; Yamamoto et al. 2006). Both RANKL and M-CSF cytokines govern several phases of the osteoclast differentiation process; RANKL is associated with osteoclast fusion and formation (the formation of mononucleated cells into the multinucleated cells), adherence of osteoclasts to target resorption site on the bone surfaces, stimulation of bone resorption since M-CSF is associated preosteoclast migration, proliferation, differentiation and survival. M-CSF is also stimulant, which induces preosteoclasts for RANKL receptor (RANK) expression and regulates osteoclasts differentiation (Swanson et al. 2006; Bellido, Plotkin, and Bruzzaniti 2014, 2019; Parvizi 2010)., RANKL and M-CSF together also promote the activation of gene expression and encourage the regulation of transcription factors in osteoclasts. The expression and secretion of RANKL are upregulated by several factors (such as vitamin D, PTH and TNF, some interleukin family members; IL-1, IL-6, IL-11), which are stimulators of osteoclastogenesis and bone resorption (Aubin and Bonnelye 2000; Piert et al. 2005). Additionally, the RANKL/RANK binding also activates different signal transduction pathways, containing the RANK intake depends on TNF receptor-associated factor 6 (TRAF6) and then TRAF6 activates several kinases signaling pathways, Jun N terminal kinase (JNK), I kappa B kinase (IKK complex) and several transcription factors signaling pathways (mitogen-associated protein kinase of 38 kDa (p38)), (c-Fos, microphthalmia-related transcription factor (MITF), activator protein-1 (AP1), nuclear factor of activated T-cell 1 (NFATc1)) for osteoclasts differentiation and function (Wong et al. 1999; Filgueira 2010; Gentili and Cancedda 2009). Especially, NFATc1 has an important role due to upregulation of the crucial genes such as tartrate-resistant acid phosphatase (TRAP), integrin β3, cathepsin K (Cat K), and MMP9. In addition, activation of M-CSF receptor (C-FMS) via M-CSF binding provides the signals for recruitment of several molecules such as Src, extracellular signal-regulated kinases (ERK) and Serine/Threonine Kinase (Akt). M-CSF stimulates osteoclasts cytoskeleton reorganization depending on a series of activation steps of Src tyrosine kinase, CBL and phosphoinositide 3-kinase (PI3K), in turn. In contrast, Akt activation promotes osteoclast survival in this signal pathway (Ross 2011; Suda 1999; Sharma et al. 2012; Wagner and Eferl 2005; Boyle, Simonet, and Lacey 2003).

There is another cytokine, which is called osteoprotegerin (OPG) as a member of TNF receptor superfamily, playing a specific role in osteoclastogenesis and bone resorption process. OPG is a secreted protein in soluble form by bone marrow stromal cells, osteoblasts and osteocytes. Due to the lack of transmembrane domain OPG has no direct signaling capability as RANK and C-FMS have. The main function of OPG in the bone-related process is inhibition of several osteoclasts' activities. The RANKL-OPG ratio and osteoclast number are directly correlated with each other. Because OPG binds the available RANKL and prevents RANKL/RANK interaction and osteoclast formation. In this way, RANKL-OPG ratio effects osteoclastogenesis and RANKL/RANK/OPG mechanism may be called "a key negotiator" for osteoclastogenesis (Boyce and Xing 2008; Phan, Xu, and Zheng 2004; Florencio-Silva et al. 2015).

Bone resorption process begins with the interaction between mature osteoclast and bone matrix basal membrane. There are two main interaction domains (the sealing zone, ruffled border) that maintain the osteoclastic bone resorption. The sealing zone, consists of actin-associated proteins (F-actin), several other proteins (vinculin and paxillin) and other structural kinases such as GTP-hydrolyzing enzymes (GTPases), protein-tyrosine kinase 2-beta/focal adhesion kinase 2 (Pyk2), etc., is the key cell-membrane binding domain for osteoclasts-extracellular mineralized matrix attachment (Akisaka et al. 2001; Bellido, Plotkin, and Bruzzaniti 2014, 2019; Filgueira 2010). Additionally, in the sealing zone, the engagement of non-collagenous bone matrix proteins (such as arginine-glycine-aspartic acid (RGD) sequences-containing proteins, osteopontin, bone sialoprotein) and osteoclast podosome belt (including CD44) containing $\alpha\gamma\beta3$ integrin manages osteoclasts polarization and determines a restriction area for the ruffled border domain so that the resorption pits (Howship's lacunae) below osteoclast are isolated (Luxenburg et al. 2007; Chabadel et al. 2007; Florencio-Silva et al. 2015). The ruffled border, formed by microvilli in order to transport lysosomal and endosomal vesicles, has also a crucial role as well as the sealing zone for osteoclastic activity. Also, the vacuolar-type proton pump (H^+/ATPase or V/ATPase), which located in the ruffled border, allows H^+ ions (composed by carbonic anhydrase) releasing for increasing acidity of the resorption pit and the chloride channel, H^+/Cl^- exchange transporter 7 (CIC-7), release Cl^- ions simultaneously with V/ATPase into the resorption pit in order to avoid intracellular polarization. Meanwhile, cell charge neutrality is sustained with bicarbonate/chloride channels on the basolateral membrane. Thus, HCl formed depending on H^+ and Cl^- releasing and dissolves inorganic part (hydroxyapatite crystals) of mineralized bone (Saltel et al. 2004; Ohno et al. 2001; Graves et al. 2008). After the dissolution of hydroxyapatite crystals, TRAP, Cat K, MMP-9, and MMP-14 secreted into the resorption cavity for degradation of the bone organic phase (Cappariello et al. 2014; Filgueira 2010; Akisaka et al. 2001). Following the entire matrix resorption, products are endocytosed with helping the ruffled border and internalized into the cell and then secreted from the basolateral membrane via transcytosis (Bellido, Plotkin, and Bruzzaniti 2014, 2019). At the end of bone resorption, mature osteoclasts allocate from targeted bone surface and undergo apoptosis.

It is necessary to mention that MMPs may play several roles in osteoclastogenesis and in bone resorption. The expression of MMPs by osteoclasts has a complex process in bone resorption. These MMPs and their expression depend on the species and the type of ossification. But, also the expression of the MMPs is directly related with several osteoclasts function (such as migration and adhesion) and several steps of bone resorption processes (especially, the initiation and termination of the bone resorption). There are many MMP types have been observed in the resorption cavities of bone, such as MMP-1, MMP-2, MMP-3, MMP-9, MMP-10, MMP-12, MMP-13, and MMP-14 (Andersen et al. 2004; Takahashi et al. 2008; Delaisse et al. 2003). The present of MMP-14 supports cell fusion, starting with preosteoclasts activation via RANKL expression and continuing with multinucleated cell formation. Moreover, some types of metalloproteinases mediate the shedding of RANKL in osteoclastogenesis such as MMP-14, MMP-7, and a disintegrin and metalloproteinase 10 (ADAM10). They have a prominent attribute, shedding of membrane-bound RANKL in suitable conditions. In addition, MMP-14 also can shed CD44, which is crucial for cell adhesion and migration for bone resorption (Hikita et al. 2006; Takahashi et al. 2008; Kusano et al. 1998). RANKL and vitamin D enable MMP-9 expression in osteoclasts where localized on poorly mineralized bone surfaces and MMP-9 degrades the unmineralized bone matrix. On the other hand, several factors (such as cytokines, ECM proteins, etc.) may affect several osteoclast functions as MMP-9 substrate in the osteoclastogenesis, for example; MMP-9 transform passive pro-TNF-α molecule in the active cytokine, which has a considerable impact for osteoclasts survival and activity. Unlike the other MMPs, MMP-13 is expressed by osteoblasts or osteoblast like-bone lining cells, which is close to mature osteoclasts, and is secreted by them for degradation of type I collagen and then recruitment of mature osteoclasts on the bone resorption surface (Wittrant et al. 2003; Crimmin et al. 1995; Vivinus-Nebot et al. 2014).

3. Matrix Metalloproteinases, General Properties and Function

MMPs or matrixins constitute a structurally and functionally associated zinc endopeptidase family. These extracellular proteinases regulate development and physiologic events such as wound healing, morphogenesis, angiogenesis and ECM degradation. MMPs degrade (*in vitro* and *in vivo*) protein components present in ECM (Bode et al. 1999; Vu and Werb 2000). These enzymes share common activation mechanisms and functional domains, they require Ca^{2+} and Zn^{2+} and have activity at neutral pH values.

MMPs are synthesized as transmembrane proenzymes and are activated by the removal of an amino-terminal propeptide. A cysteine residue present in this peptide interacts with zinc which is in the active site of the enzyme, by this way propeptide keeps the enzyme inactive. Enzyme activation is achieved by the disruption of this interaction which initiates the "cysteine switch" mechanism. MMP activation can be done by chaotropic agents or propeptide cleavage by MMP family members or other proteases. MMP family degrades most of the ECM components. MMP family members are more than 20 now. MMP subgroups are based on the preferred substrate or structural domain similarities. For example, MMP subgroup which acts on fibrillar collagen is named as collagenases and which act on denatured collagen as gelatinases (Vu and Werb 2000).

MMPs show distinct but moderately overlapping functions. This abundance is possibly for the prevention of the organism from the regulatory control loss. Since MMPs have a great potential to degrade the ECM, remodel or destroy tissues, their regulatory control is a multiple level process. MMP proteolitic activity is regulated with TIMPs, which are protein structured reversible inhibitors of MMPs. Furthermore extracellular matrix protein 1 and RECK, a membrane-attached glycoprotein are known other MMP inhibitors (Illman, Lohi, and Keski-Oja 2008).

3.1. MMP Domain Structure

All of the MMPs are synthesized as proenzymes, which are inactive, and released into the extracellular space, except membrane bound MMPs (MT-MMPs) which are attached to the plasma membrane. A common catalytic mechanism and domain structure is shared by MMPs (Figure 1.). Generally, MMPs are formed from a signal peptide, a prodomain, a catalytic domain, a hinge region, and a hemopexin domain (Löffek, Schilling, and Franzke 2011; Visse and Nagase 2003; Mecham and Parks 1998).

Signal peptide contains 17-20 residues, mostly hydrophobic amino acid residues, and works as a signal for the release in endoplasmic reticulum to be finally exported from the cell. Only MMP-17 does not contain this signal peptide (Mecham and Parks 1998).

Propeptide domain consists of approximately 80 amino acid residues and contains an extremely preserved and unique cysteine sequence, the cysteine switch, which is able to bind the zinc present in the catalytic domain resulting in enzyme deactivation. In order to activate the enzyme this cysteine-zinc bond is proteolytically cleaved and propeptide domain is removed revealing the catalytic domain. Besides calcium ions are required for the enzymatic activity (Jones, Sane, and Herrington 2003; Mecham and Parks 1998).

Furin-cleavage site insert is approximately nine residues and leads to intracellular cleavage via furin. MMP-11, MMP-14, MMP-15, MMP-16 and MMP-17 have this insert. External proteases cleave the propeptide from the middle in the rest of the enzymes. Zinc is partially exposed starting the autolytic cleavage of the remnant propeptide.

Catalytic domain contains approximately 160-170 residues and includes calcium ion and structural zinc atom binding sites. Catalytic zinc binding site is included in 50-54 residue at the C-terminal end of the catalytic domain. MMP-2 and MMP-9 contain three repeats of fibronectin type II domain (fibronectin-like repeats) in their catalytic domain. Fibronectin-like repeats assists the enzyme-gelatin substrate binding.

Catalytic domain is connected to the hemopexin domain using a linker region, which is referred as hinge region. Enzymes (e.g., MMP-7) that does not contain a hemopexin domain does not need a hinge. Hemopexin domain contains approximately 200 residues. Although it does not seem to be fundamental for catalytic activity all MMPs except MMP-7 involve this structure. Besides, it has effect on substrate specificity and TIMP binding (Mecham and Parks 1998).

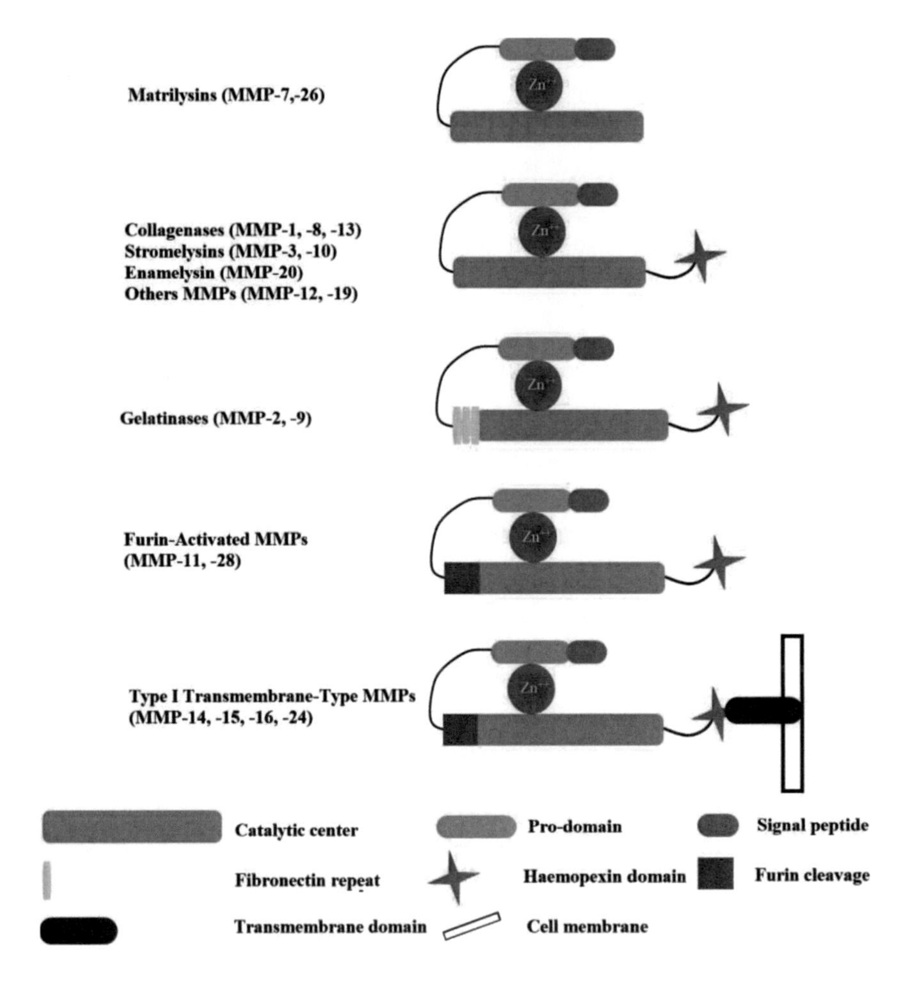

Figure 1. Domain structures of some MMP groups.

3.2. MMP Family

Collagenase, first member of MMP family, was discovered during metamorphosis in the tail of a tadpole by Jerome Gross and colleagues in 1962, which was able to degrade rigid collagen rods. This enzyme degrades triple helix of the collagen at neutral pH values with an almost particular activity (Brinckerhoff and Matrisian 2002; Jones, Sane, and Herrington 2003; Visse and Nagase 2003).

MMPs are traditionally divided into six subgroups according to their substrate specificity and structural homology. These MMPs have a common domain structure but all of the domains are not represented in the all family members (Stetler-Stevenson and Yu 2001). These subgroups are; 1) collagenases, 2), gelatinases 3) stromelysins, 4) matrilysins, 5) membrane-type MMPs and 6) others (Paiva and Granjeiro 2014).

3.2.1. Collagenases

This subgroup involves MMP-1, MMP-8, MMP-13 and MMP-18. These enzymes as a common feature have the ability to cleave the interstitial collagens I, II and III at a cleavage site, which is in three-quarters distance from the N-terminus of the substrate collagen. Collagenases can degrade several other ECM/non-ECM components (Visse and Nagase 2003) but fibrillar collagen cleavage is particularly limited to collagenases. Many other proteases, including other MMPs, can degrade the denatured collagen substrate following the higher order destruction of fibrillar collagens by collagenases (Matrisian 1992).

3.2.2. Gelatinases

This subgroup involves MMP-2 (gelatinase A) and MMP-9 (gelatinase B). Denatured collagens (gelatins) are readily degraded by these enzymes. Both of the gelatinases contain three pieces of type II fibronectin domain, which bind to gelatin, collagens and laminin, in their catalytic domain. Gelatinase B also contain a domain similar to type V collagen.

Gelatinase A degrades type I, II and III collagens, while gelatinase B do not (Visse and Nagase 2003; Matrisian 1992).

3.2.3. Stromelysins

This subgroup involves MMP-3 (stromelysin 1) and MMP-10 (stromelysin 2). These enzymes have this name due to their comparatively wide substrate specificity. Both of the enzymes possess a similar substrate specificity. However, generally stromelysin 1 has a higher proteolytic effectivity than stromelysin 2. Furthermore, stromelysin 1 activates several proMMPs, and has a critical action in the formation of totally active MMP-1 from partially processed proMMP-1. Although MMP-11 is named as stromelysin 3, it is generally presented in "other MMPs" subgroup due to the differences in substrate specificity and structure from stromeysin 1 (Visse and Nagase 2003; Matrisian 1992).

3.2.4. Matrilysins

This subgroup involves MMP-7 (matrilysin 1) and MMP-26 (matrilysin 2). Enzymes present in this subgroup are characterized by hemopexin domain deficiency. Matrilysin 1 and 2 degrade some ECM components. However matrilysin 1 also processes molecules present on cell surface (e.g., Fas-ligand, pro-α-defensin, and pro-TNF-α) (Visse and Nagase 2003). Furthermore, the structure of matrilysin 1 comprise of the propeptide and the catalytic domains, which makes it the smallest family member (Vu and Werb 2000).

3.2.5. Membrane-Type MMPs

This subgroup involves six membrane-type MMPs (MT-MMP). Four of these are type I transmembrane proteins (MMP-14 or MT1-MMP, MMP-15 or MT2-MMP, MMP-16 or MT3-MMP and MMP-24 or MT5-MMP) and two of these are glycosylphosphatidylinositol (GPI) anchored proteins (MMP-17 or MT4-MMP and MMP-25 or MT6-MMP).

All the MT-MMP enzymes, except MT4-MMP, are able to activate proMMP-2. Furthermore MT-MMPs are able to degrade several ECM components. MMP-14 has the ability to break down the type I, II and III collagens (collagenolytic activity). MMP-24 is expressed mostly in the cerebellum, it is specific to brain (Visse and Nagase 2003; Mecham and Parks 1998; Folgueras et al. 2009).

Main difference of MT-MMP from other MMPs is that it contains a membrane spanning hydrophobic sequence in the forth and last pexin-like repeat of the C-terminal domain (Birkedal-Hansen 1995). All MT-MMPs commonly contain an insertion of a potential furin/prohormone convertase cleavage site at the end of the propeptide domain (Mecham and Parks 1998).

3.2.6. Other MMPs

There are seven MMP enzymes which are not classified in the above subgroups. These are MMP-12 (metalloelastase, macrophage elastase), MMP-19, MMP-20 (enameysin), MMP-21, MMP-23 (cysteine array MMP), MMP-27, MMP-28 (epilysin) (Visse and Nagase 2003; Löffek, Schilling, and Franzke 2011)

Metalloelastase is expressed primarily in macrophages and it degrades elastin and several other proteins and plays an essential role in the migration of macrophages. Enamelysin is mainly situated inside the newly formed tooth enamel and degrades amelogenin (Visse and Nagase 2003).

MMP-21 was reported to be characterized from cDNA by Ahokas et al. in 2002 (Ahokas et al. 2002). Although they suggested that this enzyme has a function in tumour progression and embryogenesis its physiological substrates remain unidentified (Skoog et al. 2006).

MMP-23 is expressed primarily in reproductive tissues. MMP-23 prodomain does not contain the cysteine switch motif. Furthermore, the enzyme does not contain the hemopexin domain. But it contains a cysteine-rich domain which is followed by an immunoglobulin-like domain.

Since its propeptide contains a furin recognition motif, its cleavage takes place in the Golgi, and it is released into the extracellular environment as an active enzyme (Visse and Nagase 2003).

Epilysin (MMP-28) is the newest member of the matrix metalloproteinase (MMP) family of extracellular proteases. Enzyme is expressed primarily in keratinocytes. Altough it is suggested that it might have functions in the tissue hemostasis and wound repair, its functions are not known entirely (Illman, Lohi, and Keski-Oja 2008; Visse and Nagase 2003).

4. Roles of MMPs During Bone Development, Remodeling, and Repair

Degradation of the organic matrix in the bone depends on the activity of proteolytic enzymes, which consist of 2 major classes: the cysteine proteinase family (such as cathepsin K) and the MMP family. Over the past years, different mammalian MMPs have been identified (Table 1.). These MMPs are all zinc-dependent endopeptidases with the ability to degrade the organic matrix at physiological pH (Paiva and Granjeiro 2017).

The MMPs have an active role in natural skeletal development, remodelling, and repair. They have expressed and regulated in the bone, dentin and cartilage cells. A lot of information about MMPs we learn is coming from *in vivo* studies using MMP gene knockout mice, genetic diseases involving MMP gene mutations in human and preclinical experimental arthritis models (Liang et al. 2016). Stromelysins such as MMP-3 not only act as a metalloproteinase but also activate a latent proMMP. Therefore, the cooperative effects of collagenases, gelatinases, and stromelysins may be important for MMP-dependent degradation of bone matrix (Kusano et al. 1998). In this part, we will discuss each MMP and interactions of MMPs with other proteins in bone development, remodelling, and repair. The roles of important MMPs in bone development are discussed as follows.

Table 1. The characteristics of the human MMP family members

Names Collagenases	Size* (kDa)	# of Domains	ECM Substrates	Non-ECM Substaretes
MMP-1	43	4	Type I, II, III, V, IX, VII, X collagens; gelatins; aggrecan; entactin; tenascin; perlecan	Pro-IL-1β; α2M; α1-PI; α1-antichymotrypsin; CTGF
MMP-8	55	4	Type I, II, III, VII, X collagens; gelatins; aggrecan; fibronectin; fibrinogen; proteoglycans	Angiotensin; L-selectin; α1-PI
MMP-13	42	4	Type I, II, III, IV, IX, X, XIV collagens; aggrecan; fibronectin; tenascin	SDF-1; MCP-3; endostatin; α1-antichymotrypsin; CTGF; α2M; pro-TGF-β
Gelatinases				
MMP-2	62	5	Gelatins; type IV, V, VII, XI collagens; laminin; fibronectin; elastin;	Pro-IL-1β; pro-TGF-β; α1-PI; MCP-3; plasminogen; galectin-3
MMP-9	82	6	Gelatins; type III, IV, V, XI collagens; aggrecan; elastin	Pro-IL-1β; IL-2 receptor α; pro-TGF-β; α1-PI; ICAM-1; plasminogen
Stromelysins				
MMP-3	43	4	Aggrecan; gelatins; fibronectin; type II, III, IV, V, IX, X collagens; elastin; laminin	pro-IL-1β; IGF-BP-3; α1-PI; α1-antichymotrypsin; α2M plasminogen; E-cadherin;
MMP-10	44	4	Aggrecan; elastin; fibronectin; laminin; type I, III, IV, V collagens	ProMMP-1; ProMMP-8; ProMMP-10

Names	Size* (kDa)	# of Domains	ECM Substrates	Non-ECM Substaretes
MMP-11	46	4	fibronectin; laminin; aggrecan; type III, IV, IX, X collagen; gelatins	α1-PI; α2M; IGF-BP-1
Matrilysins				
MMP-7	19	3	Aggrecan; elastin; entactin; gelatins; fibronectin; laminin;	RANKL; pro-TNF; Fas-L; Pro-α-defensin; β4 integrin;
MMP-26	19	3	Gelatin; type IV collagen; fibronectin; fibrinogen	α1-PI; proMMP-9
Membrane-Anchored MMPs				
MMP-14	54	7	Type I, II, and III collagen; gelatins; aggrecan; fibronectin; laminin; fibrin	CD44; α2M; SDF-1 ProMMP-2; ProMMP-13
MMP-16	56	7	Type III collagen; fibronectin; gelatins	KISS-1; ProMMP-2; ProMMP-13
Other MMPs				
MMP-20	22	4	Amelogenin; aggrecan; gelatin;	Unknown

* Moleculer weight in active form.

α2M, α2-macroglobulin; CTGF, connective tissue growth factor; IGF-BP, insulin-like growth factor binding protein; KISS, Metastin; MCP, monocyte chemoattractant protein; PI, proteinase inhibitor; SDF, Stromal cell-derived factor; TGF, transforming growth factor.

4.1. MMP-1

MMP-1 (collagenase-1, interstitial collagenase or fibroblast collagenase) is the most predominant proteinase of the MMP family (Sofat 2009). The main substrate of the MMP-1 is type I collagen. Additionally, it may degrade types II, III, V, IX and X fibrillar collagens which are the most abundant proteins in the body (Ziober et al. 2000; Arakaki, Marques, and Santos 2009). *Mmp1* gene is located on the chromosome 11, at the 11q22.3

band, which is expressed in fibroblasts, osteoblasts, synovial cells, chondrocytes, epithelial and endothelial cells (Geng et al. 2018). MMP-1 can be inhibited by TIMP-1 and TIMP-2. MMP-1 is stimulated by TNF-α, parathyroid hormone (PTH) and IL-1 (Liao and Luo 2001). Expression of these molecules regulates the bone resorbing and forming, thus, imbalance of these ones could lead to the reduced organization of bone structure (Bord et al. 1999).

The MMP-1 has a crucial role in bone resorption and the prevention of bone loss (Liao and Luo 2001). The abnormal synthesis and activity of MMP-1 may cause some bone and cartilage-related diseases such as osteoporosis, OA. Kaspiris et al. reported that MMP-1 level was increased in OA patients (Kaspiris et al. 2013). In pathological circumstances, *Mmp1* gene expression level highly increases and thus contributes an abnormal level of connective tissue elimination (Vincenti and Brinckerhoff 2002). Multiple studies have shown that *Mmp1* gene polymorphism has a relationship with primary RA and OA in different nationalities (Geng et al. 2018; Abd-Allah et al. 2012; Luo et al. 2015). The higher expression level of the *Mmp1* gene was seen in RA and OA patients (Tetlow, Adlam, and Woolley 2001). Barlas et al. reported that *Mmp1* rs1799750 1G/2G gene polymorphism resulted in knee OA in the Turkish population (Barlas et al. 2009). Luo and colleagues' results confirm the previous studies (Luo et al. 2015), however, some studies about this gene have indicated that there is no relationship with knee OA even in the same country (Kara et al. 2016). So, *Mmp1* gene and OA relation is still a discrepancy. However, targeting MMP-1 may contribute a powerful therapeutic alternative for OA that affects not only the cartilage but the bone as well (Kaspiris et al. 2013).

The cartilage has ECM components and chondrocytes, which is mainly comprised of proteoglycans, aggrecans, and collagens. These components are rich about type II collagen. Type II collagen has a triple-helical structure which gives resistance to degradation. MMP-1, MMP-8, and MMP-13 can degrade type II collagen into gelatine. After, these gelatines are degraded by gelatinases into small peptides (Krane 2001; Liang et al. 2016). Therefore, MMP-1 is very critical for bone and cartilage development and remodelling (Vincenti et al. 1996).

MMP-1 is also found on gingival crevicular fluid and it has an important role on dental degradation. MMP-1 is localized in destructive periodontitis (Balli et al. 2016). On chronic periodontitis patients, MMP-1 level increased while the amount of TIMP-1 decreased. At the sixth week after the initial treatment, those two enzyme levels were close to the limits of healthy individuals (Tuter et al. 2005). Dahan et al. examined that MMP-1 mRNA expression in gingival biopsies taken from healthy individuals and periodontitis diagnosis patient and showed that MMP-1 mRNA expression could be detected only in periodontitis patients (Dahan et al. 2001).

4.2. MMP-2

MMP-2 (gelatinase A or CLG4A) is a 62 kDa gelatinase capable of cleaving collagen types I, IV, V, VII, and XI together with aggrecan, gelatin, fibronectin, laminin, large tenascin-C, and elastin. Mutations of the autosomal recessive gene expressing MMP-2 are found in patients with hereditary multicenter osteolysis, arthritis, and the expression of this gene encodes the 16q12-2 region (Liang et al. 2016).

MMP-2 are gelatinases that degrade collagens IV and V, elastin and gelatin components of the ECM, as well as the subendothelial basement membrane thereby increasing vessel fragility (Kaneshiro et al. 2016). Proteolytic degradation of the endothelial basement membrane and other matrix components is associated with immune cells. Macrophage-derived MMP-2 is an important regulatory proteolytic enzyme in the migration of immune cells into the brain. MMP-2 plays an important role in vascularization (Öncel 2012). MMP-2 activity is increased via the Endo180/MMP-14 axis. Native collagen fragments produced by collagenases are favoured substrates for endocytosis via uPARAP/Endo180. MMP-2 may also cleave these fragments before internalization, suggesting that uPARAP/Endo180 and MMP-2 participate in parallel steps of collagen degradation. Results when taken together, proved that MMP-14 is a collagenase which coordinates the phagocytosis of collagen (Paiva and Granjeiro 2017).

Production of ECM and degradation of bone cells are critical levels at bone metabolism. Mutation of the genes of MMP-2 enzymes characterized by subcutaneous nodules, arthropathy, and local osteolysis that human genetic defects. Defects on *MMP-2*$^{-/-}$ relates to decreasing mineral concentration of bone at the femur and increasing bone volume at calvaria. MMP-2 plays a crucial role in the formation and maintenance of the osteocytic canalicular network. Osteocytic network formation is a marker of bone remodeling and mineralization (Inoue et al. 2006).

Information on the effects of bone absorbing factors on the regulation of MMP-2 is limited, and Lorenzo et al. showed that MMP-2 expressed in osteoblasts is not regulated by bone-absorbing factors (Kusano et al. 1998)

4.3. MMP-3

MMP-3 (stromelysin-1, SL-1, STR1 and STMY1) has 43 kDa molecular weight and can cleave a broad variety of substrates including collagen types II, III, IV, V, IX, and X, non-collagenous ECM proteins including elastin, fibronectin, laminin, and aggrecan and activate latent growth factors (Paiva and Granjeiro 2017; Bonassar et al. 1995). Of the MMP, MMP-3 is a critical member because of its wide substrate specificity. Additionally, MMP-3 not only degrades the ECM proteins mentioned above but also conduces to the activation of proMMP-1 which has an active role in ECM degradation. This contribution greatly increases the efficiency of MMP-3 in matrix fragmentation (Sasahara et al. 1999).

Mmp3 gene is located on chromosome 11, at the 11q22.3 band. It is expressed and produced by fibroblasts, smooth muscle cells, synovial cells, chondrocytes, osteocytes and osteoblasts (Meikle et al. 1992; Munhoz, Godoy-Santos, and Santos 2010). Its expression is regulated by growth factors such as TGF-β, cytokines such as IL-1-β, TNF-α, tumour promoters and oncogene products (L M Matrisian 1990). Polymorphism of *Mmp3* gene (rs3025058) in the promoter region was first reported in 1995 (Ye et al. 1995). There are two common polymorphisms; 5A and 6A (5 and 6 consecutive adenines). These two polymorphisms on *Mmp3* gene causes OA

(Takahashi et al. 1999), RA (Constantin et al. 2002; Scherer et al. 2010) and ankylosing spondylitis (Wei et al. 2009).

Frzb$^{-/-}$ mice (deletion of the Frizzled-related protein (FRZB)) caused increase of articular cartilage loss during arthritis. This cartilage loss in *Frzb*$^{-/-}$ mice caused increase of MMP-3 expression and activity. Further, Frzb inhibited MMP-3 activity (Lories et al. 2007). In a study conducted by Bord et al. it was observed that MMP-3 may be required for soft tissue degradation, allowing cell migration and making space for the expanding bone during bone formation (Bord et al. 1998).

Inhibition of the *Mmp3* expression in osteoblast cells prevents the bone loss and improve bone formation in osteoporotic bone fractures (Zarka-Prost-Dumont et al. 2014; Paiva and Granjeiro 2017) Regardless of the cause, once the cartilage damage begins, OA develops as a result of a destruction process; degradation of the matrix, local inflammation and cytokine release by the release of MMP-3 (Elmali et al. 2002). Increasing the expression of *Mmp3* excessively and insufficient inhibition by TIMP-1 develop joint erosion during the development of RA (Visse and Nagase 2003). In RA patients, the MMP-3 level in serum and synovial fluid has risen. The serum level of MMP-3 is mainly originated from synovium fluids (Yoshihara et al. 1995; Catrina et al. 2002). Takahashi et al. reported that mRNA expressions level of *Mmp3* in cartilage and synovium during OA has increased. Additionally, hyaluronan could be used as a therapeutic agent for early stages of OA in the rabbit model because of that hyaluronan is upregulated MMP-3 (Takahashi et al. 1999).

Bisphosphonates inhibit activities of several MMPs by chelating Ca^{2+} and Zn^{2+} ions, which are essential for activity of these enzymes. Almost all bisphosphonates (except ibandronate) are able to inhibit MMP-3 activity (Paiva and Granjeiro 2017). Alendronate hydroxyapatite nanocomposites scaffold, one of the bisphosphonates, were seeded onto osteoblasts and osteoclasts cells. This nanocomposite decreased osteoclast proliferation while osteoblast proliferation and differentiation were enhanced. The MMP-1 and MMP-3 expression were decreased, and bone development was increased in long bones by using zoledronic (Välimäki et al. 2006; Paiva and Granjeiro 2017).

4.4. MMP-7

MMP-7 (Matrilysin-1, uterine metalloproteinase, or pump-1 protease (PUMP-1)) is expressed in epithelial cells, macrophages, osteoclasts, and monoosteophils (Krane and Inada 2008; Zhang and Shively 2013). *Mmp7* gene is located on chromosome 11, 11q22.2 band and consists of only ~10 kbp and has 19 kDa molecular weight (Knox et al. 1996). MMP-7 is firstly discovered in the rat uterus (Woessner and Taplin 1988). MMP-7 degrades a variety non-collagenous proteins of ECM substrates, including fibronectin, elastin, entactin, proteoglycans, and also casein, gelatins I, II, III, IV, and V, and activate many latent growth factors (Barillé et al. 1999; Yokoyama et al. 2008; Paiva and Granjeiro 2017). MMP-7 can cleave some cell surface molecules such as pro-α-defensin, Fas-ligand, pro–TNF-α, and E-cadherin (Loreto et al. 2013). Among other MMPs, MMP-7 is a very short amino acid sequence and is the smallest of the known MMPs (Wilson and Matrisian 1996). Additionally, the C-terminal protein domain does not exist in MMP-7 (Stamenkovic 2003). MMP-7 has not only its own proteolytic activity but also the capacity to activate other latent forms MMPs. MMP-7 could activate MMP-1 and MMP-2 and activation of these two MMPs help bone resorption. (Barillé et al. 1999; Imai et al. 1995).

Proteoglycans, glycoproteins, and adhesion molecules are important to achieve functional ECM structure for bone repair, bone resorption, coupling to bone formation, and implant surface. Together with MMP-2 and TIMP-2, MMP-7 is involved in early stages of ECM remodelling and MMP-7 play an essential important role in bone repair/regeneration (Coe et al. 2010; Shubayev et al. 2004; Paiva and Granjeiro 2017). MMP-7, like the other MMPs, removes the collagenous layer from the bone surface by breaking it into small molecules and trigger osteoclastic resorption to achieve resorption cavity (Delaissé et al. 2000). Osseointegration is integration of implants permanently to living bone. During the osseointegration, osteocytes are very active and enlarged on the implant sites (Brånemark et al. 1977; Clokie and Warshawsky 1995). Shubayev et al. investigated the change in immunoreactivity of ECM-controlling MMPs, TIMP-3 and TNF-α during the osseointegration of titanium implant to bone (Shubayev et al. 2004). The level of MMP-2, MMP-7, and MMP-9 are increased consistently (up to 12 weeks) in permanent titanium implants animal as well as TIMP-3, and TNF-α, but not observed in unoperated side of the bone. In the implanted-bone side, osteocytes were greatly immunoreactive for MMP-2 and MMP-7 at all weeks. The increased expression level of MMP-7 in osteocytes and collagen bundles of haversian canals during osseointegration may prove that MMP-7 tries to maintain ECM content which is essential for further bone regeneration. MMP-7 regulates bone remodelling during implantation via; (i) maintaining the content of proteoglycan matrix (ii) chondrocyte function at growth plate and (iii) adhesion molecules contributing to the solubilization of osteoid that is essential in maintaining bone turnover. During osseointegration, *Mmp7* expression increased by endochondral ossification in chondrocytes of the growth plate which is promoted new bone forming (Shubayev et al. 2004). During bone repair, human monoosteophils, derived from LL-37 (natural peptide) treated human $CD14^+$ monocytes, express and secrete high level of MMP-7, approximately 600 times higher than untreated controls (Zhang and Shively 2013). Overexpression of MMP-7 in osteoclasts contributes to increased osteoclast activation and may cause rheumatic osteoporosis (Yang et al. 2013). In a study conducted by Salamanna et al. increased levels of MMP-7 during osteoclastic activity in the ovariectomized (OVX) rats was observed. MMP-7 plays an important role in the evolution of bone loss in osteoporosis (Salamanna et al. 2015). Compared to healthy ones, patients with periodontal disease have higher MMP-7 level in their Gingival Crevicular Fluid (GCF). It is thought that MMP-7 is released into GCF for early defence in periodontal diseases (Şurlin et al. 2014).

MMP-7 was localized in bone-resorbing osteoclasts. MMP-7 overexpression in the tumour-bone microenvironment may be correlated with the in tumour-induced bone resorption and tumour cell invasion (Ishikawa et al. 1996). Barillé et al. reported that in multiple myeloma cancer, MMP-7 level is highly expressed and caused bone elimination and facilities the invasive behaviour of cells (Barillé et al. 1999). Another study reported that MMP-7 has profound effects on cell-cell communication. It gets soluble to RANKL which contributes progression of the prostate cancer-induced osteolysis and further supports osteoclast activation. It was shown that MMP-7 was highly expressed by osteoclasts in tumour-bone interface both *in vivo* and *in vitro* conditions during the development of prostate tumour cells (Lynch et al. 2005). Osteoclast-derived MMP-7 provides RANKL solubilization which contributes tumour invasiveness and tumour-induced osteolysis in human breast-to-bone metastases (Thiolloy et al. 2009). Wilson et al. reported that *Mmp7*$^{-/-}$ mice represented decreasing intestinal tumorigenesis in about of average tumour diameter and mean tumour multiplicity (Wilson et al. 1997). Tough *Mmp7*$^{-/-}$ mice reduced tumorigenesis, there is no difference seen in skeletal remodelling. In lymph node-negative breast carcinoma, of the 75% studied carcinoma, MMP-7 was expressed. Its overexpression may contribute to *in vitro* invasiveness of breast cancer (Chung et al. 2004).

4.5. MMP-8

MMP-8 (human neutrophil collagenase or the collagenase-2) has 75 kDa molecular weight as highly glycosylated proenzyme which is produced in polymorphonuclear leukocyte (PMN), some non-PMN-lineage cells such as periodontal ligament (PDL) fibroblasts, gingival fibroblasts and gingival epithelial cells in the oral cavity (Ingman et al. 2005; Apajalahti et al. 2003). Additionally, smooth muscle cells, chondrocytes, plasma cell, and macrophages could produce MMP-8. MMP-8 prefer to degrade type I and III collagens and also digest type VII and X collagens (Kurzepa et al. 2014). Besides, it can degrade some structural substrates such as cartilage aggrecan, proteoglycans, fibronectin, fibrinogen as well as non-structural substrates such as angiotensin, some chemokines, and L-selectin (Lint and Libert 2006).

Mmp8 gene has been expressed and regulated in skeletal tissues and cells (Krane and Inada 2008; Hanemaaijer et al. 1997). Besides, during the stage of embryonic development in the rat mandible and hind limb, bone and cartilage cells express *Mmp8* gene (Sasano et al. 2002). *Mmp8* gene is located on chromosome 11, at the 11q22.3 band and consists of ~250 kbp

(Krane and Inada 2008; Cheng et al. 2009). Sasano et al. concluded that MMP-8 expressed in the osteoblastic cell lineage, such as mesenchymal condensation, osteoblasts, and osteocytes. Therefore, MMP-8 may play an important role in the remodelling of ECM structure during embryonic development of bone and cartilage. Additionally, MMP-8 may also contribute to the bone healing process during fracture of bone (Sasano et al. 2002). Inada et al. reported that MMP-8 has functions during cartilage development in older animals. *Mmp8* and *Mmp13* null allele and double null allele were studied for investigating the role of *Mmp8* gene for cartilage development. *Mmp13*$^{-/-}$ mice show similar growth plates in embryos and new-borns while double null *Mmp8*$^{-/-}$and *Mmp13*$^{-/-}$ 6–12 week old mice represent great growth plate disturbances (Inada et al. 2006).

MMP-8 is produced from dentin-forming odontoblasts (Sulkala et al. 2007). MMP-8 is the major collagenase of MMP in dentin, which degrades ECM in periodontal disease such as tooth decay (Balli et al. 2016). Studies represented that MMP-8 is related to articular cartilage turnover and OA (Li et al. 2011; Rose and Kooyman 2016; Sulkala et al. 2007). Although bone is similar to dentin in organic and mineral content, mineralized dentin is not remodeled in a similar way to bone. Sulkala et al. showed that in the dentin organic matrix organization, MMP-8 is highly expressed especially before mineralization step (Sulkala et al. 2007). Gelatinase activity is richer in the mineralized organic matrix of dentin than in the non-mineralized dentin. Thus, it would seem that relatively more gelatinases than MMP-8 are organized or more efficiently/actively bound to the mineralized dentin matrix. This suggests that MMP-8 may be primarily related to the organization of the dentin organic matrix before. The mineralized dentin is not subject to a substantial reorganization, however, predentin and non-mineralized dentin structure inside the dentinal tubules are probably organized in response to the functional demands and maintenance of odontoblast attachment. Some altering in intratubular collagen matrix content may result in decreasing of the dentin mechanical properties such as flexure strength and resistance to fatigue (Wang et al. 2013; Sulkala et al. 2007).

Since MMPs are key mediators in the destruction of periodontal tissues, their presence in periodontal tissues and GCF have been investigated by many researchers to date. The latent form of MMP-8 was found to be associated with gingivitis, whereas active MMP-8 was responsible for tissue destruction in periodontitis (Romanelli et al. 1999; Teles et al. 2010). In addition, it has been reported that patients with periodontitis and peri-implantitis are MMP-8 and MMP-9 originating from peritoneal polymorphonuclear neutrophils (PMNL) as the dominant MMP in the GCF (Golub, Ryan, and Williams 1998; Golub et al. 1994). The GCF and gingival collagenase activity in individuals with periodontitis have been reported to be higher than healthy individuals (Sorsa et al. 1999; Nomura et al. 2008). Kraft-Neumarker et al. studied that the GCP samples were taken from patients with generalized chronic periodontitis and determined a significant relationship between active MMP-8 levels and periodontal pocket depth (Kraft-Neumärker et al. 2012). In periodontitis patients with advanced attachment loss, increased collagenase activity (MMP-8) in GCP is an important finding proving that collagenase is involved in tissue destruction (Lee et al. 1995). Mäntylä et al. investigated three different groups; healthy, gingivitis and periodontitis patients. GCF MMP-8 levels in areas with severe periodontitis were >1 mg/L and were differentiated from healthy and gingivitis areas with a positive test result. It has been reported that an increase or change in the level of MMP-8 may be beneficial in the maintenance of periodontal therapy (Mäntylä et al. 2003).

4.6. MMP-9

MMP-9 (type IV collagenase or gelatinase B) is 82 kDa gelatinase in same group with MMP-2. MMP-9 has an important role in connective tissue remodeling and basement membrane transformation (Malemud 2006). It has high specific activity in the degradation of denatured collagen in ECM. MMP-9 can break down elastin, wild type IV, V, XI collagen and non-ECM molecules such as amyloid beta peptide, substance P, and myelin basic protein. However, MMP-9 cannot degrade proteoglycan, unnatural type I collagen, or laminins. The expression of MMP-9 varies according to the development stages. In early development, MMP-9 is expressed in trophoblasts, osteoclasts, indicates bone resorption and a role in implantation. Mature MMP-9 is expressed in inflamed cells during various cancer types (Liang et al. 2016).

MMP-9 is thought to have many methods affecting bone development. Long bones in *Mmp9*$^{-/-}$ mice are 10% shorter than wild type mice. MMP-9 is thought to have many methods affecting bone development. Long bones in MMP-9 knockout mice are 10% shorter than wild type mice. The activity of MMP-9 in the growth plate during endochondral bone formation plays a role in influencing the structural properties of all bones. All MMP-9 expressing critical cells are of bone marrow origin (Liang et al. 2016) MMP-9 is found in monocyte-macrophages and osteoclasts, but not in osteoblastic cells (Kusano et al. 1998).

There are multiple mechanisms by which MMP-9 may influence skeletal cell differentiation during bone repair by mechanical stimulation in periosteal cells and inflammatory cells. MMP-9 plays important roles in bone development and repair. These enzymes are involved in the interaction between skeletal progenitors and inflammatory cells. MMP-9 has been assessed with inflammation and the differentiation of skeletal progenitors under different mechanical stimuli at the fracture site. Wang et al. studied on MMP-9 to operate the bone remodeling and the inflammatory phases. MMP-9 can regulate skeletal cell fate and distinguish chondrocytes in stabilized and non-stabilized fractures in the meanwhile inflammatory process. MMP-9, which can regulate the mechanical properties of ECM and can directly affect periosteal stem cells, has several indirect mechanisms of action. MMP-9 indirectly enhances the bioavailability of VEGF, controls the distribution of secreting cells and thus regulates angiogenesis (Wang et al. 2013).

In a study by Bruni-Cardoso et al., prostate-derived bone metastases, osteolytic and osteogenic responses, osteoclasts, proteinase, the main source of MMP-9 have been revealed. The effect of host-derived MMP-9 on prostate tumour progression in bone was determined. Host MMP-9 has been shown to significantly contribute to the growth of prostate tumour without affecting the osteolytic or osteogenic change caused by the prostate tumour. Osteoclast-derived MMP-9 contributes to angiogenesis without altering osteolytic or osteogenic changes induced by prostate tumour and promotes prostate tumour growth in the bone microenvironment. The analysis of the amount of MMP-9 during bone destruction revealed no difference in the number of osteoclasts present at the tumor-bone interface. In a study examining multiple myeloma progression, MMP-9 has been reported to play an important role in myeloma-induced bone destruction. MMP-9 also controls the bioavailability of VEGF-A164. According to the analysis at the mRNA level, VEGF-A164 expression indicates that MMP-9 expresses VEGF-A164 at similar levels in osteoclast cultures. MMP-9 appears to be an important tumour in the bone microenvironment, partly due to the angiogenesis of VEGF-A164. In osteoclast angiogenesis and potent anti-osteoclast therapy in osteoclasts, close to osteoclasts and endothelial cells in areas of osteoclasts, e.g., VEGF-A, bone remodeling, several proangiogenic factors of osteoclasts, angiogenesis and potent anti-osteoclast treatments have been demonstrated (Bruni-Cardoso et al. 2010).

4.7. MMP-11

MMP-11 (Stromelysin-3, SL-3, ST3, or STMY3) is expressed in osteoblastic cells and FGF-2 and TGF-β regulate its expression level (Varghese 2006). *Mmp11* gene is located on chromosome 22, at the 22q11.23 band and consists only ~16 kbp (Levy et al. 1992). MMP-11 has 46 kDa molecular weight and can cleave a broad variety of substrates including collagen types III, IV, IX, and X, non-collagenous ECM proteins including laminin, elastin, entactin, proteoglycan, fibronectin, aggrecan and gelatins in normal physiological processes (Galis and Khatri 2002). In contrast to other MMP's, it weakly degrades ECM structural proteins (Waresijiang et al. 2016). However, MMP-11 plays an essential role in ECM modelling by regulating the release of growth factors from ECM proteins (Masson et al. 1998). MMP-11 is intracytoplasmically produced and released as an active enzyme and it can be activated by furin, a family of endopeptidases, inside the cell (Curran and Murray 2000).

MMP-11 is responsible for ECM degradation and bone remodelling as well as invasion and metastasis of carcinomas. It may be highly expressed and produced locally in bone loosening area. Between bone and prosthesis of aseptic loose artificial hip joints, expression of level MMP-2, MMP-7, MMP-8, MMP-11, and MMP-14 was increased moderately in synovium-like interface tissues (Takei et al. 2000). Konttinen et al. reported that MMP-11 was highly expressed in synovial tissue of trauma and RA patients (Konttinen et al. 1999). MMP-11 expression was more obvious in the human postnatal growth plate. Of the stromelysins, MMP-11 was found to be dominant in chondrocytes, osteoblasts, and osteocytes in growth plates (Haeusler et al. 2005). Decreasing level of proteolysis due to down-regulation of MMP-11 may be concluded with the accumulation of cartilage and bone. Reduction of MMP-11 mRNA level resulted in avian tibial dyschondroplasia lesions (Velada et al. 2011). During osteocytogenesis prosses in the MLO-A5 cells (late osteoblastic cell line), MMP-11 mRNA level is increased whose is related to mineralization onset (Prideaux et al. 2015). MMP-11 regulates ECM structure for the acquisition of the osteocyte phenotype during bone remodelling.

Overexpression of the MMP-11 has been associated with bone metastasis and tumour progression in many studies. Nonsrijun et al. study represent that high expression of MMP-11 was reported in prostatic adenocarcinoma (Nonsrijun et al. 2013). Expression of the MMP-11 in prostate cancer patients has shown the parallel result with positive-bone metastasis (Roscilli et al. 2014). MMP-11 does not only encourages tumour progression by improving invasion migration, survival, and cancer cells but also modify tumour microenvironment via break down ECM components. Besides, MMP-11 performs potent antitumoral impact on solid tumours by suppressing metastasis (Zhang et al. 2016). Eiró et al. demonstrated that overexpression of MMP-11 was obtained in breast cancer patients. MMP-11 expresses a lot of different proinflammatory cytokines, which conducts to bone metastasis in mononuclear inflammatory cells (MICs) positive patients (Eiró et al. 2013).

4.8. MMP-13

MMP-13 (collagenase-3, or CLG3) is a member of the family of proteolytic enzymes which is involved in the breakdown of cartilage ECM components (Fu et al. 2016). MMP-13 is originated from osteoblasts, chondrocytes, and synovial cells (Aiken and Khokha 2010). MMP-13 is a major extracellular collagenase and can be efficiently activated by urokinase plasminogen activation (Paiva and Granjeiro 2017). Efficiently reduces collagen type II. MMP-13 disrupts the ECM components all along physiological development and also in pathological conditions such as tumour invasion, OA, and metastasis. The major functional features of MMP-13 are directly associated to the ability to degrade interstitial collagen, a critical structural component of all connective tissues, including bone, and other matrix-related components (Fu et al. 2016).

MMP-13 is expressed in hypertrophic chondrocytes and osteoblasts. It endorses the removal of hypertrophic cartilage from the growth plate and restructuring of newly accumulated trabecular bone during long bone development. MMP-13 is directly effective during angiogenesis in the early stages of cartilage ECM degradation. The increase in activity of MMP-13 is directly related to cartilage degradation. Clinical studies have shown that patients with articular cartilage destruction have high *Mmp13* expression. Furthermore, *Mmp13*, which is overexpressed in transgenic mice, develops spontaneous OA-like articular cartilage destruction phenotype (Nakatani, Chen, and Partridge 2016). Fu et al. reported that MMP-13 derived from multiple myeloma serves as a strong secretagogue of osteoclast fusion and bone-absorbing activity, regardless of proteolytic activity. MMP-13 has been shown to play a non-protective role in the development of lytic bone lesions in multiple myeloma patients (Fu et al. 2016).

Tang et al. examined the role of MMP-13 in the remodeling and maintenance of bone matrix and subsequently in fracture resistance using *Mmp13* deficient mice. They observed high non-enzymatic cross-linking and hypermineralization, collagen disorder and concentric regions of canalicular malformation throughout the diaphysis of MMP-13 deficient tibia. The results of the researchers have shown that the middle cortical bone matrix has an important role in MMP-13 in osteocyte perilacunar remodeling and is necessary for maintaining bone quality (Tang et al. 2012). uPARAP/Endo180 has been described as a receptor for MMP-13. Binding of MMP-13 would cause deterioration of the functionality through the low-density lipoprotein receptor related protein. uPARAP/Endo180 and MMP-13 are co-expressed during endochondral and intramembranous ossification. They have also been reported to have potential interaction during bone development (Paiva and Granjeiro 2017; Inoue et al. 2006). Attur et al. investigated the effect on MMP-13 by using disintegrin ADAMTS-4 to reduce, regulate periostin levels, and examine the effect of type II collagen. They have observed that the dose of MMP-13 expression and periosty significantly increase the expression of ADAMTS-4 mRNA and increase cartilage degeneration by collagen and proteoglycan degradation. Periost induction of MMP-13 expression has been shown to be inhibited by CCT031374 hydrobromide, an inhibitor of the canonical Wnt/y-catenin signaling pathway. In addition, siRNA-mediated theft of endogenous periostin has shown that it blocks structural MMP-13 expression. Results of study showed that periostin is a catabolic protein that promotes cartilage degeneration by upregulating MMP-13 with canonical Wnt signals (Attur et al. 2015). Nakatani et al. showed that *Hdac4* suppresses *Mmp13* expression by inhibiting the activity of *Runx2* and that *Hdac4*$^{-/-}$ mice show increased levels of MMP-13 mRNA and protein in hypertrophic chondrocytes and trabecular bone. They found that elevation of MMP-13 contributed to the phenotype of *Hdac4*$^{-/-}$ mice (Nakatani, Chen, and Partridge 2016).

4.9. MMP-14

MMP-14 (membrane-type 1 MMP, MT1-MMP) is a cell membrane-bound protein in perisceletal and skeletal tissue. It is an important collagenase that plays a role *in vitro* preosteoblast differentiation, acting on cell shape. *Mmp14* knockout cells undergoing osteogenic differentiation do not differentiate into osteoblasts and adipogenic and chondrogenic differentiation. MMP-14 can degrade type I, II, and III collagen, gelatins, aggrecan, fibronectin, laminin and fibrin (Okada 2017; Krane 2001).

MMP-14 is effective in regulating cell morphology. MMP-14 is a major intracellular collagenase in all bone cells and is responsible for introducing collagen fragments for endocytosis and lysosomal degradation with the uPARAP/Endo180 receptor. MMP-14 osteoclast migration is critical for monocyte-macrophage fusion (Paiva and Granjeiro 2017).

MMP-14 has been characterized in human bone pathologies and skeletal syndromes. In MMP-14 deficiency, apoptosis balance is disrupted in chondrocyte and cartilage disruption. Normal endochondral ossification is prevented. Therefore, MMP-14 has an important role in bone development and growth (Pullen et al. 2018). MMP-14 is important in osteoblast/ osteocyte transition control during TGF-β activation. MMP-14 maintains osteoblast survival when osteoblasts pause synthesizing new bone matrix, maintaining osteoblast-osteocyte transition, and inhibiting osteoclast apoptosis. At this time, physiological mineralization and hydroxyapatite formation, which is controlled by ECM, occurs. Inorganic phosphate (Pi) and inorganic pyrophosphate (PPi) regulate molecules such as alkaline phosphatases that are not specific to a certain tissue, regulate the formation of hydroxyapatite and contributes to the increase in the required Pi levels. The osteoblast mechanism controlled by MMP-14 is important for bone regeneration (Nakano, Addison, and Kaartinen 2007).

MMPs and in particular MMP-14 mediate tumour progression and tissue injury by activation of TGF-β. MMP activity also increases cell migration. Identification of CD44 by MMP-14, for example, increases cell activity in different cancer cell cultures. MMPs also have an important role in vascularization. MMP-2, MMP-9, and MMP-14 play a major role in tumour vascularization (Öncel 2012). Co-expression of CD44 and MMP-14 promotes increased CD44 activity and cell migration. Here, CD44 acts as a coupling molecule, i.e., a "trap receptor" to maintain active MMP-9 form on the cell surface, and aids in the activation of proMMP-9 in the osteoclast cell membrane (Paiva and Granjeiro 2017).

4.10. MMP-16

MMP-16 (membrane-type 3 MMP, MT3-MMP or MT-MMP3) is expressed and presented in different parts of the bone structure including epiphysis, periosteum, chondrocytes, osteoblasts, osteocytes and primary ossification center in the bone during development of the bone (Shi et al. 2008; Aiken and Khokha 2010; Liang et al. 2016). MMP-16 has 56 kDa molecular weight in activated form and *Mmp16* gene is located on chromosome 8, at 8q21.3 band and it consists of ~300 kbp (Tatti et al. 2011). MMP-16 can degrade type III collagen, fibronectin, and gelatin thus promoting bone growth and development (Okada 2017). Besides, it has a very important role aside from the degradation of ECM components for bone remodelling: activation of the proMMP-2 by cleavage (Haeusler et al. 2005).

In a study, the results showed that MMP-16 with MMP-2, MMP-3, MMP-9, and MMP-13 had an important role in bone and cartilage turnover, providing accessing of blood into the growth plate and cartilage replacement by newly formed bone tissue during the endochondral ossification in turkey. MMP-16 was expressed in the epiphysis, in the compact bone, in the reserve zone of the growth plate, in cells surrounding the blood vessels penetrating the hypertrophic zone during ossification process (Simsa, Genina, and Ornan 2007). Overexpression of MMP-16 brings about higher degradation of aggrecan and collagen and decreasing cellularity, thus causing dehydration and degeneration of discs which makes a significant contribution to the formation of intervertebral disc degeneration (IDD). The level of MMP-16 mRNA and protein, as well as the level of MMP-2 in IDD patient, was higher than healthy (non-degenerated cadavers) discs (Zhang et al. 2017). MMP-16 was expressed moderately during the bone remodeling in the bone-implant interface of the hip joint implant (Takei et al. 2000)

The precise functions of mesenchymal cells which are expressed on skeletal tissue provide degradation of ECM components with MMP-16. *Mmp16*$^{-/-}$ mice has represented decreasing viability and migration of mesenchymal cells which caused by the lack of cleavage of fibrillar collagen, leading to dwarf bone growth. *Mmp14*$^{-/-}$ and *Mmp16*$^{-/-}$ double knockout mouse have resulted in great developmental bone abnormalities, leads most of the mouse death in the first day of birth while *Mmp16*$^{-/-}$ mice did not show any death in the first day of the birth. This result depiced that post-embryonic bone development in mice does not depend on the presence of the MMP-16. *Mmp14*$^{-/-}$ and *Mmp16*$^{-/-}$ double knockout mouse is not only causing the death of mouse but also represent some bone abnormalities including shorter, thin cortical bone, decreased bone mineralization, poorly formed bones and domed skulls (Shi et al. 2008). Additionally, *Mmp14*$^{-/-}$ and *Mmp16*$^{-/-}$ double knockout mouse also affects cartilage development. This mouse has expanded hypertrophic zone in the long bones and significantly impaired bone formation at the primary ossification centre. This depicts that thc role of MMP-16 is incontrovertible in the bone as well as cartilage remodelling because of reducing of mesenchymal cell involvement in skeletal structure.

4.11. MMP-20

MMP-20 (enamel metalloproteinase or enamelysin) is expressed by epithelial ameloblasts and mesenchymal odontoblasts. *Mmp20* gene is located on chromosome 11, 11q22.2 band and consists only ~49 kbp (Wang et al. 2013). MMP-20 provides degradation of amelogenin, the largest structural component of the enamel matrix, and plays an important role in enamel development thanks to its structural and enzymatic properties (Bartlett et al. 2004). Tooth-forming cells are known to secrete huge amounts of MMP-20 (Erkli and Ersöz 2011). The MMP-20 activity or expression and following dentin formation could be mediated by dentin-derived growth factors such as TGF-β1. The TGF-β1 induces moderately downregulation of MMP-20 expression (Palosaari et al. 2003).

MMP-20 could cleave enamel matrix proteins such as amelogenin (Amel), ameloblastin (Ambn) and enamelin (Enam) (Guo et al. 2015). MMP-20 can be secreted into the dentinal fluid by natural human odontoblasts. MMP-20, which is produced during the milk dentition period and participates in the structure of dentin, is released during the decay development process (Sulkala et al. 2002). Another enzyme related to enamel formation is MMP-2. However, MMP-20 is more important than MMP-2 in the formation of the appropriate enamel formation (Caterina et al. 2002). A study with *Mmp20*$^{-/-}$ mouse showed that the amelogenin was not processed properly and there were changes in the enamel matrix. The resulting enamel has a structure that is thinner than the normal, and have a distorted prismatic structure (Caterina et al. 2002). Barlett et al. observed that the enamel mineral content of the *Mmp20*$^{-/-}$ mice was approximately equal to half of that of healthy mice, and enamel hardness was observed to have a ratio of 2 to 3 compared to healthy cells (Bartlett et al. 2004). Mutations on *Mmp20* gene results in autosomal-recessive hypomaturation amelogenesis imperfecta (Kim et al. 2005). In *Mmp20* null mice, the enamel matrix develops as a bilayer without rod-interrod boundaries organization (Caterina et al. 2002). Guo et al. reported that on Bone morphogenetic protein 2 (*Bmp2)* conditional knock out mice, the expression level of *Mmp20* is reduced as reflected in a reduced enamel formation in the tooth (Guo et al. 2015). Salmela et al. reported that exposure of the tributyltin, which is pollutant from seafood and known risk for bone related disease, results in decreasing of the *Mmp20* expression in odontoblasts. The decreasing expression of *Mmp20* expression by tributyltin exposure cause impair enamel mineralization, delay of cell differentiation and dentin mineralization (Salmela et al. 2012).

The dental fluorosis is a defect in enamel mineralization. It was caused by increased porosity due to the high amount of *fluoride* exposure during tooth development and was firstly introduced by Black and McKay in 1916 (Black and McKay 1916; Fejerskov, Thylsrup, and Larsen 1977; Hannas et al. 2007; Erkli and Ersöz 2011). Exposure to large amounts of *fluoride* during enamel formation suspends the removal of amelogenes that need to be hydrolysed by proteinases. The reason for this delay in protein hydrolysis; it may be through MMPs responsible for the degradation of enamel proteins (Hannas et al. 2007). *In vitro* micromolar concentrations of *fluorid*e have been observed to cause changes in metalloproteinase activity. In addition, it was observed that *fluoride* causes decrease in amelogenin hydrolysis by MMP-20. The *in vivo* study depicted that fluoride uptake could change the relative amount of active MMP-20 in mature enamel (DenBesten et al. 2002).

5. Role of MMP Inhibitors in Bone Modelling, Remodelling, and Resorption

MMPs have a pivotal role in several physiological processes like embryogenesis, stem cell differentiation, normal tissue remodeling, cell motility, wound healing, apoptosis, and angiogenesis (Visse and Nagase 2003; Paiva and Granjeiro 2017; Liang et al. 2016). Normally the functions of MMPs are strictly controlled and regulated by various intrinsic inhibitors including; tissue inhibitors of metalloproteinases (TIMPs) and other endogenous protein inhibitors, such as α2-macroglobulin (α2M), the reversion-inducing cysteine-rich protein with Kazal motifs (RECK protein), thrombospondins (Chang, Hung, and Chang 2008), tissue-factor-pathway-inhibitor 2 (TFPI2), and the procollagen C-terminal proteinase enhancer (CT-PCPE) (Maskos and Bode 2003; Bedi et al. 2010; Nagase, Visse, and Murphy 2006). Besides their intrinsic inhibitors, the degenerative activity of MMPs is also controlled by gene expression of MMPs, stimulation by cytokines and growth factors, secretion and activation of proMMPs (Takaishi et al. 2008; Cawston 1996). In case of an imbalance between MMPs and their inhibitors, diseases such as atheroma, arthritis, aneurysms, cancer, RA, OA, osteoporosis, and fibrosis occur (Liang et al. 2016).

The TIMPs are natural, endogenous inhibitors currently comprised of four variants (TIMP-1, -2, -3, -4) with low molecular weights ranging between 21 (TIMP-2) and 27.5 (TIMP-1) kDa (Bord et al. 1999a; Bedi et al. 2010). These four members have sequence homology ranging from 41 to 52%. Despite the similarities each member display different surface structural features thus each has different specificity (Löffek, Schilling, and Franzke 2011; Maskos and Bode 2003). TIMPs bind MMPs noncovalently in a 1:1 molar ratio and inhibit active MMPs relatively with low selectivity (Gomis-R¨th et al. 1997; Bord et al. 1999). TIMPs inhibit MMPs by preventing substrate binding to MMPs' catalytic domain via their N-terminal domains, the flexibility of enzymes' binding interface decreases and causes steric hindrance (Grossman et al. 2010; Nagase, Visse, and Murphy 2006; Bord et al. 1999).

TIMP-1, TIMP-2, and TIMP-4 are secreted proteins, however, TIMP-3 is bound by connective tissue matrix and shows activity where it was formed (Paiva and Granjeiro 2017; Bord et al. 1999). Besides their inhibitory functions, TIMPs have some MMP independent biological roles like promoting cell growth, antiapoptotic (Stetler-Stevenson 2008) and antiangiogenic activity (Maskos and Bode 2003), stimulating embryogenesis, suppressing the growth of vascular endothelial cells, etc. (Murate and Hayakawa 1999; Nagase, Visse, and Murphy 2006). Gasson et al. reported the erythroid potentiating activity of TIMP-1 (Gasson et al. 1985). Amongst 4 members of TIMPs, TIMP-1 and TIMP-2 are most abundantly expressed during bone development and have effects on bone resorption (Dew et al. 2000; Liang et al. 2016). The bone tissue remodelling cycle has three main stages which are resorption, reversal, and formation. There is strict coordination between resorption and formation stages to provide the structural integrity of the bone. Resoption is the stage that old bone matrix digested by osteoclasts (Kenkre and Bassett 2018). TIMP-1 and TIMP-2 show a concentration-dependent effect on osteoclastic bone resorption. The stimulative effect of TIMP-1 and TIMP-2 on osteoclastic bone resorption in fetal calf serum (Shibutani et al. 1999) and rabbit mature osteoclasts (Sobue et al. 2001) was reported at concentrations lower than 50 ng/ml. It was also reported that TIMP-1 and TIMP-2 show inhibitory effects at high concentrations (Dew et al. 2000). Since other inhibitors used in experiments could not mimic the stimulative effect of TIMPs, this effect could not directly be associated with the TIMPs' inhibitory effect on MMPs. In a study conducted by Liang et al. that inhibition effect of TIMP-1 on human bone marrow-derived mesenchymal stem cells (hBMSCs)' proliferation and osteogenic differentiation were investigated. The effect of TIMP-1 on the physiological functions of hBMSC was evaluated by generating TIMP-1 stable overexpressing hBMSCs and TIMP-1 stable knockdown hBMSCs. The results showed that, compared to control groups, TIMP-1 knockdown groups remarkably increased the hBMSC growth and

TIMP-1 overexpression significantly decreased the up-regulation of osteocalcin during osteogenic differentiation, whereas TIMP-1 knockdown made a counter effect. Additionally, reduction in deposition of calcium nodules was observed when TIMP-1 was overexpressed (Liang et al. 2019).

Besides the stimulative and inhibitory effects on bone resorption, TIMPs also contribute to the regulation of bone development and remodelling. Geoffroy et al. found that TIMP-1 overexpression in mice osteoblasts caused an increase of trabecular bone volume and a decrease in bone turnover (Geoffroy et al. 2004). In another study, the location of TIMP-1 and MMPs in pathological and healthy bone tissues (neonatal rib, osteophytes and heterotopic bone) was investigated by using specific antibodies. It was reported that the expression of MMP in pathological tissues was similar to that of healthy tissues, but there were significant differences in TIMP-1 expression. The presence of TIMP-1 in healthy tissues has shown that it has an important role in the regulation of bone formation possibly due to MMP inhibitory activity (Bord et al. 1999).

The activities of chondrocytes are crucial for the growth of longitudinal bone. Cellular proliferation and hypertrophy are two main stages of a chondrocyte life cycle before bone formation. Matrix mineralization occurs during hypertrophy stage, which is the last stage of chondrocyte differentiation (Hunziker 1994; Shum et al. 2003). In a study conducted by Poulet et al., the effect of overexpressed TIMP-3 on adult bone and growth plate was investigated by generating a transgenic mouse model. The results of the histological analysis showed that when TIMP-3 was overexpressed in mouse chondrocytes the width of the growth plate significantly reduced (~40%) and hypertrophic and proliferating zones decreased remarkably. Furthermore, when compared to wild type (WT) lower numbers of osteoblasts and decreased bone mass was also observed in transgenic mouse. These findings also support the idea that TIMPs contribute to the regulation of bone formation (Poulet et al. 2016).

Osteoarthritis (OA) is a chronic disease which progressively degenerates articular cartilage. Injuries and chronic overuse are the main causes of the disease and it mainly affects the elderly. It is fairly common in society and when left untreated can lead to disability, therefore, early diagnosis is very important. Age, gender, obesity, and exercise can be counted among the risk factors of the disease (Haq, Murphy, and Dacre 2003). After the observation of highly increased MMP levels during OA, TIMP-MMP imbalance thought to be the reason for cartilage degradation during the process. (Dean et al. 1989). Based on these findings TIMP-1 levels in synovial fluids were assessed as a potential biomarker for the development of hip OA. As a result 600 ng/ml found to be critical value to predict the progression of the disease and recommended to be counted amongst risk factors of hip OA (Chevalier et al. 2001).

Tumour growth, tumour-cell invasion, and metastasis can be counted among diseases related to MMP-TIMP imbalance, on this basis possible therapeutic effects of TIMPs on cancer tissues are widely investigated (Maskos and Bode 2003). There are several studies on the use of selective MMP inhibitors in the treatment of bone metastasis caused by breast and prostate cancer which are supported by independent preclinical studies. Treatment with BB-94 or GM6001, an MMP inhibitor, has been shown to inhibit tumour growth and tumour-induced osteolysis. However, there are no studies examining the effect of MMP inhibitors on osteogenic bone metastases. For clinical use of MMP inhibitors, selective inhibition of metalloproteinases is a prerequisite to avoid the described disadvantages of the original MMP inhibitors. Fine-tuning the specificity of MMP inhibitors may be an appropriate approach in the development of therapies for the treatment of bone metastases. SB-3CT, an MMP inhibitor with high selectivity for MMP-2 and MMP-9, has been shown to be effective in inhibiting PC-3 prostate tumour progression with reduced tumour growth, vascularization, and osteolysis (Jia et al. 2014).

The RECK protein is the only membrane-anchored MMP inhibitor. The *RECK* gene was first isolated in 1989 by using a technique which has been designed to isolate cDNAs inducing flat reversion in a v-Ki-ras-transformed mice fibroblasts' NIH3T3 cell line (Takahashi et al. 1998). RECK protein is a 110 kDa molecular weight glycoprotein with various protease inhibitor-like domains and multiple epidermal growth factor-like (EGF-like) repeats. Although the *RECK* gene is expressed in various normal human organs it's mostly undetectable in tumour tissues and downregulated by oncogenes such as rat sarcoma oncogene (RAS). RECK has been reported to negatively regulate MMP activities at various levels (Cavagis et al. 2009; Oh et al. 2001; Paiva and Granjeiro 2017). The RECK has been reported to have pivotal roles during bone remodelling as well. In a study conducted by Cavagis et al. in order to investigate the effect of MMP inhibitors on bone remodelling, MC3T3-E1 pre-osteoblast cells were kept in osteogenic conditions for 28 days and the expression of TIMP-2 and RECK was examined. Starting from the second week of the cell culturing, downregulation of RECK and upregulation of TIMP-2 was observed (Cavagis et al. 2009).

The destruction of physical barriers around cancer tissue is crucial for the progression of cancer and the acquisition of malignant characters. MMPs exhibit proteolytic activity, leading to degradation of the ECM and thus to local growth of cancer tissue as well as invasion of cancer cells into surrounding vascular tissues. MMP-9 is known to be a key enzyme during tumour invasion and metastasis (Jeon et al. 2011; Gialeli, Theocharis, and Karamanos 2011). The restored RECK expression in tumour derived cells has been reported to restrict the invasive activity of cancer cells by inhibiting the proteolytic activity of MMP-9 (C. Takahashi et al. 1998). MMP-2 and MMP-14 are also known to negatively regulate MMP activities (Cavagis et al. 2009).

Alpha-2 macroglobulin (α2M) is a large plasma protein that inhibits MMPs with an uncommon mechanism. It has a region called 'bait' consisting of 35 amino acids. When this region degraded by proteases, it undergoes some conformational changes which activate thiol ester bonds in the structure. Proteases covalently bind to α2M via these bonds and they form a complex which can be detected by macrophage receptors and subsequently cleaned from the system. This mechanism is activated when TIMPs do not function properly (Mocchegiani et al. 2004). The potential of α2M to be a biomarker for diagnosis and monitoring progression of various diseases such as RA and osteonecrosis of the femoral head has been investigated.

The increased amount of α2M/MMP complexes and higher serum levels of α2M was observed which indicates that α2M has a potential to serve as a biomarker for these disease (Ghale-Noie and Hassani 2018; Tchetverikov et al. 2003).

6. Role of MMPs in Atypical Bone Tissue Formation

During ossification, bone forming, and remodeling are associated with chondrocytes, osteoblasts, osteoclasts, and bone marrow cells. Since these cells also provide bone regeneration in pathological conditions, it is quite important that these cells remain functionally stable to prevent the occurrence of bone diseases. MMPs contribute to homeostasis, regeneration and tissue repair during bone development by their cleaving activities in ECM and cellular interactions in the matrix-rich skeleton. However, these enzymes have also been implicated in the formation of bone-related diseases such as RA, OA, and osteoporosis due to their deficiency, overexpression or gene mutations (Paiva and Granjeiro 2017; Aiken and Khokha 2010; Liang et al. 2016).

MMP-2 is very effective in the activity and proliferation of osteoblast and osteoclast cells during bone development and remodeling. It is also known that these enzymes mediate to maintain the bone mineral density and integrity. Bone mineralization occurs via exchanging mineral ions in the channeled network in the structure of ECM. Studies have shown that in the absence of MMP-2, there are some defects such as decreased bone integrity, delayed growth, and osteopenia due to defects in the channeled network (Krane and Inada 2008; Liang et al. 2016). Besides, MMP-2 related diseases such as multicentric osteolysis and arthritis syndrome have been identified as associated with inactivation of MMP-2 expression in the human genome due to mutation (Martignetti et al. 2001; Liang et al. 2016).

MMP-9 is highly active in degradation of collagen in ECM which influence the structure, bending strength and toughness of bone by its activities during endochondral ossification. It was observed that in the circumstance without this enzyme, the bone had a brittle structure due to specific defects (Stamenkovic 2003; Liang et al. 2016). During the primer ossification center formation, the lack of MMP-9 enzyme delays the proper remodeling and enlargement of ossification due to reduced angiogenesis (Aiken and Khokha 2010). To overcome bone related disease such as RA and cancer, MMP-9 is expressed by inflammatory cells (Liang et al. 2016).

MMP-13 is considered to be the most important collagenase for the degradation of cartilage by its digestion ability of type II collagen. Spondyloepimetaphyseal dysplasia, which is defined by abnormal growth of the bone in childhood, is a genetic bone disease related to mutations in *Mmp13*. In the absence of MMP-13 enzyme, the irregular proliferation of chondrocytes is observed in the developing bones. The studies showed that the lack of MMP-13 and MMP-9 at the same time caused the more delayed formation of the primary ossification center, which also related the late formation of secondary ossification centers during bone development (Aiken and Khokha 2010; Liang et al. 2016). The lack of MMP-14 has caused abnormal cranial bone formation, short snouts, hypertelorism, and dome-shaped skulls *in vivo* mice models (Holmbeck et al. 1999). Symptoms in the *Mmp14* deficiency are similar to abnormalities in the lack of *Mmp2* and also causes decreasing the activity of MMP-2. In the lack of MMP-2 and MMP-14, model mice die shortly after birth due to the lack of these two enzymes at the same time (Liang et al. 2016).

MMP-16 is an essential enzyme for arranging bone growth and development and proper operating of mesenchymal cells expressed on skeletal tissue. The deficiency of *Mmp14* and *Mmp16* gene caused to craniofacial deformities, cortical bone shortening and revealed greater nuclei apoptosis in the bone-lining cells due to their activity of inhibited chondrocyte proliferation and cartilage remodeling (Aiken and Khokha 2010).

TIMPs are important for bone development and growth as well as regulating MMPs activities. Abnormal expressions of these inhibitors have been found to cause certain defects during bone formation. Decreased bone turnover as a result of overexpression of TIMP-1 has shown that it has important functions in inhibiting MMP activity during bone formation and remodeling. TIMP-1 and TIMP-2 have the ability to induce the bone-resorbing activity of osteoclasts independently of their inactivation activity of MMPs. Formation of OA is associated overexpression of TIMP-3 and TIMP-4 which are expressed from bone and joint tissues. Similarly, depending on the deficiencies of these inhibitors; in the absence of TIMP-3, cartilage degradation was observed due to an increase in type 2 collagen degradation (Liang et al. 2016).

In addition to MMPs and TIMPs, some of the proteins that these enzymes interact with may also exhibit atypical developments in the bone. Increased TGF-β signal, which is activated by MMP-2 and MMP-9 play a significant role in the structure and composition of bone matrix formation, leads to a decrease in bone stiffness and modulus. Bone morphogenetic proteins (BMPs) are multifunctional TGF-secreting growth factors and similar abnormalities occur during bone formation and absorption due to their deficiencies and their atypical activities on MMPs. BMP-induced Wnt/β-catenin signaling, which plays pivotal roles in various development and cell regeneration processes, is known to regulate the expression of TGF-β, MMP-2, MMP-9, MMP-13, and TIMP-1 from many bone cells and their irregular activities are associated with the progression of bone diseases such as RA and OA (Liang et al. 2016). VEGF has an important angiogenic factor, it is often expressed by endothelial cells and also synthesized by osteoblasts and OA chondrocytes in the bone. Some of the activity of this factor is regulated by MMP-9, and however, VEGF and its receptors (VEGFR) expressed by chondrocytes are associated with the OA process (Pufe et al. 2004; Paiva and Granjeiro 2017). Activated protein C (aPC), which is important for the activity of anticoagulant, directly activates MMP-2 in human umbilical vein endothelial cells. In RA, aPC inhibits inflammatory signaling by inhibiting MMP-9 in synovial fibroblasts and monocytes, via cooperation with MMP-2. This leads to cartilage disruption, which is potentially causing bone loss (Liang et al. 2016). Some diseases related to abnormal MMP activities are given in Table 2.

Table 2. Diseases caused by abnormal MMP activities

MMP Members	Potential Abnormal Activities of MMPs	Potential Disease Related to MMPs Abnormal Developments
MMP-2	Inactivation of gene expressions, Reduced bone mineralization	Osteolysis syndromes, Human nodulosis, arthropathy, Arthritis Cranio-facial abnormalities, Osteoporosis
MMP-9	Mutations, Delayed vascularization and apoptosis	Osteoporosis Rheumatoid arthritis, Focal brain ischemia
MMP-13	Mutations Delayed ossification	Spondyloepimetaphyseal dysplasia
MMP-14	Apoptosis of chondrocytes and cartilage breakdown, Preventing normal endochondral ossification	Abnormal cranial bone formation, Hypertelorism, Dome-shaped skulls
MMP-16	Inhibition of chondrocyte proliferation and cartilage remodeling	Craniofacial deformities

6.1. Bone Diseases Associated to Abnormal MMPs Activity

6.1.1. Osteoporosis

Bone remodeling is a process in which bone is turned over, modifications and repairings in the bone structure are performed, the coordination of bone formation and resorption by osteoblasts and osteoclasts achieved respectively (Brunetti, Di Benedetto, and Mori 2014; Feng et al. 2016). MMPs play a pivotal role in regulating the whole process by providing communication with main cells of the bone. However, the disruption of the balance between bone resorption and deposition leads to arise of diseases with reduced bone density and integrity, such as osteopenia and osteoporosis. The interruption of this process due to MMP deficiency, considering their activities in the ECM degradation, causes the abnormal bone development which leads to the occurrence of osteoporosis, loss of bone mass and integrity (Aiken and Khokha 2010; Feng et al. 2016). Osteoporosis is frequently associated with an increased risk of fractures, and osteoclasts serve a key role in the development of the disease (Zheng et al. 2018).

High level of MMP-2 activity is found in the circulation of osteoporotic patients. The potential role of MMP-2 in the pathogenesis of osteoporosis is related to B7-H3 which is a type I transmembrane glycoproteins which promotes osteoblastic differentiation and bone formation. Increased expressions of MMP-2 is induced by elevated B7-H3 in serum and a reduction B7-H3 in osteoblasts in osteoporotic patients. The B7-H3 deficient cells showed impaired osteogenic differentiation, leading to decreased osteopontin, and mineralized bone formation (Paiva and Granjeiro 2017). Glucocorticoids are important compounds that increase the bone resorption and decrease bone mechanical strength due to osteoblast-osteoclast apoptosis and consequently caused to the formation of osteoporosis. In a mouse model of glucocorticoid-induced (using prednisolone) osteoporosis and osteocytic osteolysis it was observed that MMP-2, MMP-9, and MMP-13 were upregulated and localized in the trabecular bone of the metaphysis. Furthermore, MMP-2 and MMP-13 were observed in the enlarged lacunae of osteocytes with condensed nuclei in the cortical bone diaphysis (Paiva and Granjeiro 2017; Feng et al. 2016). Osteoporosis is a common occurrence in chronic obstructive pulmonary disease (COPD) due to elevated MMP-9 activities (Bolton et al. 2009; Paiva and Granjeiro 2017).

6.1.2. Osteoarthritis and Rheumatoid Arthritis

Osteoarthritis (OA) is the most common bone disease and characterized by the progressive degeneration of articular cartilage. Although OA is thought to develop together with age, recent observations have shown that this disease may develop due to the early loss of bone and cartilage following high levels of bone remodeling (Burr and Gallant 2012). MMP members are found also related to OA progression. The overexpressed MMP-13 is associated with articular cartilage degeneration and joint pathology typical of OA and that MMP-13 expression is critical for OA disease progression due to its function as an ECM degrading enzyme (Li et al. 2017;Liang et al. 2016). The bone mass is reduced due to the greater bone resorption and decreased bone formation in case of MMP-14 absence. Intrinsic deficiencies to the osteogenic cells cause the abnormal activities of MMP-14 such as apoptosis of chondrocytes and cartilage breakdown preventing normal endochondral ossification. Due to the absence of this enzyme, various skeletal defects such as osteopenia, arthritis and fibrosis of soft tissues occur as well as defective endochondral ossification. During the development of hypertrophic cartilage, the formation of vascular channels that allow to ossification does not be formed in the absence of the MMP-14 enzyme. Delayed ossification, irregularities in the growth plate and dwarfism emerge following by the absence of vascular canalization. MMP-14 also exhibits a highly regulatory behavior on the development of bone cells, and in the lack of the enzyme, impairments are raised such as decreased osteogenesis, differentiation or dysfunction of osteoblasts, and an increase in osteoclasts (Burrage, Mix, and Brinckerhoff 2006; Stamenkovic 2003; Liang et al. 2016).

Rheumatoid Arthritis (RA) is an autoimmune disorder caused by the attack of the immune system cells on the joint tissues (Itoh 2017). Because of the inflamed and hyperplastic synovial membrane, the cartilage becomes damaged which is a target tissue in RA (Pap and Korb-Pap 2015). MMPs enable the activation and infiltration of inflammatory cytokines such as TNF-α, IL-1, and IL-6, which have quite important functions in the development of RA. It is known that various MMPs are produced by RA synovial tissue (Itoh 2017). Among these enzymes, MMP-12, which has an important role in the development of the diseases, is overexpressed by macrophages, causes excessive macrophage infiltration into the synovial tissue and consequently the development of arthritis (Wang et al. 2004; Itoh 2017). MMP-13 attends to invasion into cartilage by stimulation with various inflammatory proteins and contributes to the development of RA (Itoh 2017). Although there are known differences between OA and RA, the degradation of ECM by MMPs during the development of diseases is similar in both cases (Pap and Korb-Pap 2015).

6.1.3. Cancer and Angiogenesis

In some cases, overexpression of the MMPs disrupts the balance between osteoblast and osteoclasts, which regulates bone formation and resorption during bone development, leading to the reprogramming of osteoclasts by tumour cells and consequently the formation of osteolysis and tumour growth (Paiva and Granjeiro 2017). MMPs are known to play an important role in regulating cell growth and survival in the early stages of tumour formation. At this stage, MMPs are altering the structure of the essential components of the ECM to produce signals to activate growth factors such as TGF-α and Insulin-like growth factor (IGF) which support the growth and proliferation of the tumour cell. While MMPs are associated with cancer development and progression via promoting tumour growth, invasion, and metastasis, whereas natural and synthetic MMP inhibitors have been shown to reduce and even block the tumour development. However, it was suggested that TIMPs may have other functions than their MMP blocking ability. TIMPs can induce VEGF expression and promote tumor angiogenesis, tumor cell survival, and growth under some circumstances (Stamenkovic 2003).

The role of MMPs in signaling pathways of metastasis modulate the cell activities (Kessenbrock, Plaks, and Werb 2010; Paiva and Granjeiro 2017). For example; MMP-7 activates the receptor activator for RANKL, which activates osteoclasts; MMP-3, MMP-7, and MMP-9 cleave to the VEGF; MMP-3 and 10 cause vascular destabilization and MMP-2 and 9 activate TGF-β which is a cytokine involved in metastatic progression. In this way, MMPs not only cause metastasis but also play a role in the dissociation of vertebral integrity which leads to tumour progression within the spine (Paiva and Granjeiro 2017). MMPs-related osteolysis causes a variety of symptoms such as bone pain, pathological fractures, and spinal instability. In relation to these symptoms, patients with spinal metastasis have a remarkably reduced quality of their life (Liu et al. 2016).

Different activities are observed due to the balance between MMPs and TIMPs released by different stromal cell types. However, as well as balance is important for their activity, the cellular source of MMPs have important consequences on their function and activity (Kessenbrock, Plaks, and Werb

2010). MMPs are produced mainly by stromal cells in the area surrounding the tumour however it has been found that MMPs can also be expressed by tumour cells from various tissues (Paiva and Granjeiro 2017). Whereas some MMPs such as MMP-7 is primarily expressed by tumour cells, others, including MMP-2, MMP-3, and MMP-9, are expressed by stromal cells, sometimes predominantly so. Although it was initially believed that tumour-derived MMPs play the principal role in tumour progression, recent evidence from transgenic mouse models indicates that stromal cell-derived MMPs may play an equally important role. Expression of MMPs by stromal cells may be induced by tumour cell infiltration, by direct cell-cell contact, by tumour-derived growth factors, or as a result of stimulation by growth factors released from degraded ECM (Stamenkovic 2003).

In patients with breast cancer, MMP-2, MMP-7, MMP-9, and MMP-14 were accommodated with tumor size and lymph node metastasis. Furthermore, MMP-7 and MMP-14 were link with histological type. In primary and metastatic breast cancer patients, overexpression of MMP-1 and other bone metastasis-associated genes have been determined. Under TGF-β treatment the MMP-13 expression from osteoblasts is associated with bone resorption. Since osteoclasts do not express MMP-13, its action on bone osteolysis is an osteoclast dependent activity and it is regulated by TGF-β/RUNX2 pathway in osteoblasts and is a decent tumor-induced osteolysis biomarker (Paiva and Granjeiro 2017).

Several mediators for the survival of tumour cells are also regulated by MMPs. It has been suggested at least two mechanism support that MMP-7 promotes the survival of carcinoma cell rather than invasiveness. In the first approach, MMP-7 inhibits the coordination between Fas ligand and its receptor whose effective in initiating cell death by cleaving Fas ligand from the cell surface. In the second approach, MMP-7 separates the heparin-binding EGF precursor from the cell surface, as consequence apoptosis preventive signals are produced. Elevated MMP-9 and MMP-11 expressions are also effective in promoting tumour cell survival, metastasis and reducing apoptosis in tumour xenografts. MMP-9 promotes the survival of tumour cells by stimulating the activation of latent TGF-β (Stamenkovic 2003).

MMPs were also found to regulate angiogenesis due to their effectiveness in ECM degradation. MMPs play a pivotal role in angiogenesis has been provided by the deficiency of *Mmp9* gene, which displays a delay in endochondral bone formation that has been attributed to delayed neovascularization, delayed apoptosis, and ossification of hypertrophic chondrocytes (Vu et al. 1998; Stamenkovic 2003). MMP-9 has been shown to increase VEGF bioavailability and contribute to the formation of capillary tubes *in vitro* by enabling TGF-β activation. Similarly, the decrease in MMP-2 and MMP-14 expressions in tumour cells decreased angiogenesis *in vitro* models and slowed cell growth. Fibrin matrix surrounding the newly formed blood vessels is degraded by MMP-14, potentially facilitating endothelial cell penetration of tumor tissue. Furthermore, MMP-14 inhibition blocks the migration of endothelial cell and capillary tube formation *in vitro* (Stamenkovic 2003).

Although MMPs extracellular proteolysis activities are mostly related to cancer formation, angiogenesis, and inflammatory response, they are also known to act as tumour suppressors in many cases and mediate biological effects in their surrounding tissue. MMPs play an important role by facilitating the uptake of inflammatory cells by altering the function of chemokines and the bioavailability of proinflammatory cytokines. These immune regulatory functions of MMPs may be useful for cancer patients. For example, by inhibiting TNF-α activation by TIMPs, the inflammatory environment in the tumour microenvironment is reduced, thereby reducing the role of TNF-α in cancer progression (Kessenbrock, Plaks, and Werb 2010).

7. Recent Biomaterials and Bioengineering Approaches in Bone Modeling, Remodeling

New material approaches for the formation of bone tissue needed due to any loss of bone tissue have always been an area of interest in tissue engineering, bioengineering, and biomaterials. Knowledge of all molecules involved in bone formation mechanism is important for the osteogenesis as well as for the creation of new approaches, with respect to bone tissue engineering (Paiva and Granjeiro 2017). In the last few years, increasing research and knowledge on the composition, physicochemical properties, and trafficking mechanisms of each tissue enabled the development of biomimetic ECM matrices for use in the field of bone bioengineering (Kim et al. 2016; Hubmacher and Apte 2013).

MMPs, among cell secretomes, have become potential molecules to improve tissue bioengineering thanks to their critical functions during standard bone development (Paiva and Granjeiro 2017). Current research within bone tissue bioengineering aims to both develop an ECM mimicking bone tissue and provide time-dependent control of remodeling processes involved in ECM degradation and synthesis *in vitro* and *in vivo* (Paiva and Granjeiro 2017). In this context, it is important to understand how the new ECM and MMPs/inhibitors are built by transduction, gene transcription, protein translation, and molecule secretion and how biomaterial and/or accumulated ECM can be cleaved by MMPs *in vitro* and *in vivo* to develop bone repair/regeneration (Paiva and Granjeiro 2017).

MMPs are important components in many biological and pathological processes due to their ability to disrupt ECM components. These proteins bring new light to the interaction between ECM and its catabolism (Visse and Nagase 2003). There have been important researches for understanding the biochemical and structural properties of MMPs. Still, there are important issues that need to be explored and uncovered. Although there are some aspects understood related to the structure of the proMMP-2–TIMP-2 complex, the precise molecular assembly in time and space during cell migration is yet to be investigated. Future studies are important to effectively understand metalloproteinase family (ADAM and ADAMTs) (Visse and Nagase 2003).

Although the design of synthetic inhibitors of some matrixins was created by structural analysis and applied in animal models of cancer and arthritis, clinical studies have shown that they do not provide significant benefit. Such inconsistencies may have been due to the application of these treatments to patients at an advanced stage of the disease.

Other reasons are that the inhibitors are insufficient to inhibit the target enzymes in the tissue or that the non-target enzymes are inhibited. Currently, biological functions of metalloproteinases known in humans are not clearly understood (Visse and Nagase 2003). An important future challenge about MMPs is the design of specific MMP inhibitors which are beneficial for understanding the biological roles of MMPs and also improving of therapeutic treatments for diseases related to the uncompensated ECM degradation (Visse and Nagase 2003).

In a study, the evidence that MMPs can produce new small bioactive molecules from native ECM (is named by matrikines and matricryptins) (Ricard-Blum and Salza 2014) has given new interest to these enzymes (Paiva and Granjeiro 2017). MMPs and TIMPs are secreted by undifferentiated human MSCs that may play a role in the control of pericellular proteolysis. There are few studies on proteolytic cell secretome and therefore more study is needed (Lozito et al. 2014; Lozito and Tuan 2011; Paiva and Granjeiro 2017)

The epithelial-mesenchymal interactions that determine tooth morphogenesis, as well as the remodeling of bone and lamina propria during the eruptive process, require the action of enzymes capable of modifying the microenvironment of the ECM. The involvement of MMPs in tooth morphogenesis and eruptive processes has not yet been fully determined. In a study, *Mmp2* expression was evaluated in the tooth germ and surrounding tissues during the eruptive process. As a result, it was found that MMP-2 is a very valuable for remodeling of ECM and play a role in the extensive tissue remodeling during the intra- and extra-osseous phases of the tooth-eruption process (Sandoval et al. 2019). In a study about the importance of MMPs in dentistry, it has been underlined that MMPs can play a vital role in various oral biological processes (Prado et al. 2016). Although studies on MMPs secreted during tooth eruption are important developments for the treatment of tooth and jaw bone disorders, it is obvious that there are more issues to be studied.

References

Abd-Allah, Somia H., Sally M. Shalaby, Heba F. Pasha, Amal S. El-Shal, and Amany M. Abou El-Saoud. 2012. "Variation of Matrix Metalloproteinase 1 and 3 Haplotypes and Their Serum Levels in Patients with Rheumatoid Arthritis and Osteoarthritis." *Genetic Testing and Molecular Biomarkers* 16 (1): 15–20. https://doi.org/10.1089/gtmb.2011.0003

Ahokas, Katja, Jouko Lohi, Hannes Lohi, Outi Elomaa, Marja-Liisa Karjalainen-Lindsberg, Juha Kere, and Ulpu Saarialho-Kere. 2002. "Matrix Metalloproteinase-21, the Human Orthologue for XMMP, Is Expressed during Fetal Development and in Cancer." *Gene* 301 (1–2): 31–41. https://doi.org/10.1016/S0378-1119(02)01088-0.

Aiken, Alison, and Rama Khokha. 2010. "Unraveling Metalloproteinase Function in Skeletal Biology and Disease Using Genetically Altered Mice." *Biochimica et Biophysica Acta (BBA) - Molecular Cell Research* 1803 (1): 121–32. https://doi.org/10.1016/J.BBAMCR.2009.07.002.

Akisaka, T., H. Yoshida, S. Inoue, and K. Shimizu. 2001. "Organization of Cytoskeletal F-Actin, G-Actin, and Gelsolin in the Adhesion Structures in Cultured Osteoclast." *Journal of Bone and Mineral Research* 16 (7): 1248–55. https://doi.org/10.1359/jbmr.2001. 16.7.1248.

Andersen, Thomas L., Maria Del Carmen Ovejero, Tove Kirkegaard, Thomas Lenhard, Niels T. Foged, and Jean Marie Delaissé. 2004. "A Scrutiny of Matrix Metalloproteinases in Osteoclasts: Evidence for Heterogeneity and for the Presence of MMPs Synthesized by Other Cells." *Bone* 35 (5): 1107–19. https://doi.org/10.1016/j.bone.2004.06.019

Anderson, H. Clarke. 2003. "Matrix Vesicles and Calcification." *Current Rheumatology Reports* 5 (3): 222–26.

Apajalahti, S., T. Sorsa, S. Railavo, and T. Ingman. 2003. "The *in Vivo* Levels of Matrix Metalloproteinase-1 and -8 in Gingival Crevicular Fluid during Initial Orthodontic Tooth Movement." *Journal of Dental*

Research 82 (12): 1018–22. https://doi.org/10.1177/154405910308201216.

Arakaki, P. A., M. R. Marques, and M. C. L. G. Santos. 2009. "MMP-1 Polymorphism and Its Relationship to Pathological Processes." *Journal of Biosciences* 34 (2): 313–20. https://doi.org/10.1007/s12038-009-0035-1.

Arana-Chavez, V. E., A. M. Soares, and E. Katchburian. 1995. "Junctions between Early Developing Osteoblasts of Rat Calvaria as Revealed by Freeze-Fracture and Ultrathin Section Electron Microscopy." *Archives of Histology and Cytology* 58 (3): 285–92.

Aszódi, A,. J. F. Bateman, E. Gustafsson, R. Boot-Handford, and R. Fässler. 2000. "Mammalian Skeletogenesis and Extracellular Matrix: What Can We Learn from Knockout Mice?" *Cell Structure and Function* 25 (2): 73–84.

Attur, Mukundan, Qing Yang, Kohei Shimada, Yuki Tachida, Hiroyuki Nagase, Paolo Mignatti, Lauren Statman, Glyn Palmer, Thorsten Kirsch, Frank Beier, and Steven B. Abramson. 2015. "Elevated Expression of Periostin in Human Osteoarthritic Cartilage and Its Potential Role in Matrix Degradation via Matrix Metalloproteinase-13." *FASEB Journal : Official Publication of the Federation of American Societies for Experimental Biology* 29 (10): 4107–21. https://doi.org/10.1096/fj.15-272427.

Aubin, J. E., and E. Bonnelye. 2000. "Osteoprotegerin and Its Ligand: A New Paradigm for Regulation of Osteoclastogenesis and Bone Resorption." *Osteoporosis International* 11 (11): 905–13. https://doi.org/10.1007/s001980070028.

Aubin, J. E., and F. Liu. 1996. The Osteoblasts Lineage. En: Bilezikian J. P., Raisz L. G., Rodan G. A., Eds. *Principles of Bone Biology*. San Diego, California: Academic Press.

Balli, Umut, Burcu Ozkan Cetinkaya, Gonca Cayir Keles, Zeynep Pinar Keles, Sevki Guler, Mehtap Unlu Sogut, and Zuleyha Erisgin. 2016. "Assessment of MMP-1, MMP-8 and TIMP-2 in Experimental Periodontitis Treated with Kaempferol." *Journal of Periodontal & Implant Science* 46 (2): 84. https://doi.org/10.5051/jpis.2016.46.2.84

Barillé, S., R. Bataille, M. J. Rapp, J. L. Harousseau, and M. Amiot. 1999. "Production of Metalloproteinase-7 (Matrilysin) by Human Myeloma Cells and Its Potential Involvement in Metalloproteinase-2 Activation." *Journal of Immunology (Baltimore, Md. : 1950)* 163 (10): 5723–28.

Barlas, I. Ömer, Melek Sezgin, M. Emin Erdal, Günsah Sahin, Handan Camdeviren Ankarali, Zühal Mert Altintas, and Ebru Türkmen. 2009. "Association of (−1,607) 1G/2G Polymorphism of Matrix Metalloproteinase-1 Gene with Knee Osteoarthritis in the Turkish Population (Knee Osteoarthritis and MMPs Gene Polymorphisms)." *Rheumatology International* 29 (4): 383–88. https://doi.org/10.1007/s00296-008-0705-6.

Bartlett, J. D., E. Beniash, D. H. Lee, and C. E. Smith. 2004. "Decreased Mineral Content in MMP-20 Null Mouse Enamel Is Prominent During the Maturation Stage." *Journal of Dental Research* 83 (12): 909–13. https://doi.org/10.1177/154405910408301204.

Bedi, Asheesh, David Kovacevic, Carolyn Hettrich, Lawrence V Gulotta, John R Ehteshami, Russell F Warren, and Scott A Rodeo. 2010. "The Effect of Matrix Metalloproteinase Inhibition on Tendon-to-Bone Healing in a Rotator Cuff Repair Model." *Journal of Shoulder and Elbow Surgery* 19 (3): 384–91. https://doi.org/10.1016/j.jse.2009.07.010.

Bellido, Teresita, Lilian I Plotkin, and Angela Bruzzaniti. 2014. "Bone Cells." In *Basic and Applied Bone Biology*, 27–45. Elsevier.

Bellido, Teresita, Lilian I Plotkin, and Angela Bruzzaniti. 2019. "Bone Cells." In *Basic and Applied Bone Biology*, 37–55. Elsevier.

Birkedal-Hansen, Henning. 1995. "Proteolytic Remodeling of Extracellular Matrix." *Current Opinion in Cell Biology* 7 (5): 728–35. https://doi.org/10.1016/0955-0674(95)80116-2.

Black, Greene Vardiman, and F. McKay. 1916. "Mottled Teeth: An Endemic Developmental Imperfection of the Enamel of the Teeth Heretofore Unknown in the Literature of Dentistry." *Dental Cosmos* 58: 129–56.

Bode, W., C. Fernandez-Catalan, H. Tschesche, F. Grams, H. Nagase, K. Maskos, H. Nagase, and K. Maskos. 1999. "Structural Properties of

Matrix Metalloproteinases." *Cellular and Molecular Life Sciences (CMLS)* 55 (4): 639–52. https://doi.org/10.1007/s000180050320.

Boivin, G., Y. Bala, A. Doublier, D. Farlay, L. G. Ste-Marie, P. J. Meunier, and P. D. Delmas. 2008. "The Role of Mineralization and Organic Matrix in the Microhardness of Bone Tissue from Controls and Osteoporotic Patients." *Bone* 43 (3): 532–38. https://doi.org/10.1016/j.bone.2008.05.024.

Boivin, G., and P. J. Meunier. 2002. "The Degree of Mineralization of Bone Tissue Measured by Computerized Quantitative Contact Microradiography." *Calcified Tissue International* 70 (6): 503–11. https://doi.org/10.1007/s00223-001-2048-0.

Bolton, C. E., M. D. Stone, P. H. Edwards, J. M. Duckers, W. D. Evans, and D. J. Shale. 2009. "Circulating Matrix Metalloproteinase-9 and Osteoporosis in Patients with Chronic Obstructive Pulmonary Disease." *Chronic Respiratory Disease* 6 (2): 81–87. https://doi.org/10.1177/1479972309103131.

Bonassar, Lawrence J., Eliot H. Frank, Jane C. Murray, Claribel G. Paguio, Vernon L. Moore, Michael W. Lark, John D. Sandy, Jiann-Jiu Wu, David R. Eyre, and Alan J. Grodzinsky. 1995. "Changes in Cartilage Composition and Physical Properties Due to Stromelysin Degradation." *Arthritis & Rheumatism* 38 (2): 173–83. https://doi.org/10.1002/art.1780380205.

Bonewald, Lynda F. 2007. "Osteocyte Messages from a Bony Tomb." *Cell Metabolism* 5 (6): 410–11. https://doi.org/10.1016/j.cmet.2007.05.008.

Bonjour, Jean-Philippe. 2011. "Calcium and Phosphate: A Duet of Ions Playing for Bone Health." *Journal of the American College of Nutrition* 30 (5 Suppl 1): 438S-48S. http://www.ncbi.nlm.nih.gov/pubmed/22081690.

Bonnans, Caroline, Jonathan Chou, and Zena Werb. 2014. "Remodelling the Extracellular Matrix in Development and Disease." *Nature Reviews. Molecular Cell Biology* 15 (12): 786–801. https://doi.org/10.1038/nrm3904.

Bord, S., A. Horner, C. A. Beeton, R. M. Hembry, and J. E. Compston. 1999. "Tissue Inhibitor of Matrix Metalloproteinase-1 (TIMP-1) Distribution in Normal and Pathological Human" *Bone* 24 (3): 229–35. http://www.ncbi.nlm.nih.gov/pubmed/10071915.

Bord, S., A. Horner, R. M. Hembry, and J. E. Compston. 1998. "Stromelysin-1 (MMP-3) and Stromelysin-2 (MMP-10) Expression in Developing Human Bone: Potential Roles in Skeletal Development." *Bone* 23 (1): 7–12. https://doi.org/10.1016/S8756-3282(98)00064-7.

Boskey, A. L., and A. S. Posner. 1984. "Bone Structure, Composition, and Mineralization." *The Orthopedic Clinics of North America* 15 (4): 597–612.

Boyce, Brendon. F., D. E. Hughes, K. R. Wright, Lianping Xing, and A. Dai. 1999. "Recent Advances in Bone Biology Provide Insight into the Pathogenesis of Bone Diseases." *Laboratory Investigation; a Journal of Technical Methods and Pathology* 79 (2): 83–94.

Boyce, Brendan F., and Lianping Xing. 2008. "Functions of RANKL/RANK/OPG in Bone Modeling and Remodeling." *Archives of Biochemistry and Biophysics* 473 (2): 139–46. https://doi.org/ 10.1016/j.abb.2008.03.018.

Boyle, William J., W. Scott Simonet, and David L. Lacey. 2003. "Osteoclast Differentiation and Activation." *Nature* 423 (6937): 337.

Brånemark, P I, B O Hansson, R Adell, U Breine, J Lindström, O Hallén, and A Ohman. 1977. "Osseointegrated Implants in the Treatment of the Edentulous Jaw. Experience from a 10-Year Period." *Scandinavian Journal of Plastic and Reconstructive Surgery. Supplementum* 16: 1–132. http://www.ncbi.nlm.nih.gov/pubmed/356184.

Brinckerhoff, Constance E., and Lynn M. Matrisian. 2002. "Matrix Metalloproteinases: A Tail of a Frog That Became a Prince." *Nature Reviews Molecular Cell Biology* 3 (3): 207–14. https://doi.org/10. 1038/nrm763.

Brunetti, Giacomina, Adriana Di Benedetto, and Giorgio Mori. 2014. "Bone Remodeling." *Imaging of Prosthetic Joints: A Combined Radiological and Clinical Perspective* 396: 27–37. https://doi.org/10.1007/ 978-88-470-5483-7_3.

Bruni-Cardoso, A., L. C. Johnson, R. L. Vessella, T. E. Peterson, and C. C. Lynch. 2010. "Osteoclast-Derived Matrix Metalloproteinase-9 Directly

Affects Angiogenesis in the Prostate Tumor-Bone Microenvironment." *Molecular Cancer Research* 8 (4): 459–70. https://doi.org/10.1158/1541-7786.MCR-09-0445.

Buckwalter, J. A., and R. R. Cooper. 1987. "Bone Structure and Function." *Instructional Course Lectures* 36: 27–48.

Burger, E. H., and J. Klein-Nulend. 1999. "Mechanotransduction in Bone--Role of the Lacuno-Canalicular Network." *FASEB Journal : Official Publication of the Federation of American Societies for Experimental Biology* 13 Suppl (9001): S101-12.

Burr, David B. 2019. "Bone Morphology and Organization." In *Basic and Applied Bone Biology*, 3–26. Elsevier. https://doi.org/10.1016/B978-0-12-813259-3.00001-4.

Burr, David B., and Maxime A. Gallant. 2012. "Bone Remodelling in Osteoarthritis." *Nature Reviews Rheumatology* 8 (11): 665–73. https://doi.org/10.1038/nrrheum.2012.130.

Burrage, Peter S, Kimberlee S. Mix, and Constance E. Brinckerhoff. 2006. "Matrix Metalloproteinases: Role in Arthritis." *Frontiers in Bioscience : A Journal and Virtual Library* 11 (1): 529-43.

Cappariello, Alfredo, Antonio Maurizi, Vimal Veeriah, and Anna Teti. 2014. "Reprint of: The Great Beauty of the Osteoclast." *Archives of Biochemistry and Biophysics* 561: 13–21. https://doi.org/10.1016/j.abb.2014.08.009.

Capulli, Mattia, Riccardo Paone, and Nadia Rucci. 2014. "Osteoblast and Osteocyte: Games without Frontiers." *Archives of Biochemistry and Biophysics* 561: 3–12.

Caterina, John J, Ziedonis Skobe, Joanne Shi, Yanli Ding, James P Simmer, Henning Birkedal-Hansen, and John D Bartlett. 2002. "Enamelysin (Matrix Metalloproteinase 20)-Deficient Mice Display an Amelogenesis Imperfecta Phenotype." *The Journal of Biological Chemistry* 277 (51): 49598–604. https://doi.org/10.1074/jbc.M209100200.

Catrina, A. I., J. Lampa, S. Ernestam, E. af Klint, J. Bratt, L. Klareskog, and A-K Ulfgren. 2002. "Anti-Tumour Necrosis Factor (TNF)-Alpha Therapy (Etanercept) down-Regulates Serum Matrix Metalloproteinase

(MMP)-3 and MMP-1 in Rheumatoid Arthritis." *Rheumatology* 41 (5): 484–89. https://doi.org/10.1093/rheumatology/41.5.484.

Zambuzzi, Willian F., Claudia L. Yano, Alexandre D. M. Cavagis, Maikel P. Peppelenbosch, José Mauro Granjeiro, Carmen V. Ferreira. 2009. "Ascorbate-Induced Osteoblast Differentiation Recruits Distinct MMP-Inhibitors : RECK and TIMP-2," *Molecular and Cellular Biochemistry* 322 (1-2): 143-150. https://doi.org/10.1007/s11010-008-9951-x.

Cawston, Timothy Edward, 1996. "Metalloproteinase Inhibitors and the Prevention of Connective Tissue Breakdown." *Pharmacology and Therapeutics* 70 (3): 163–82. https://doi.org/10.1016/0163-7258(96) 00015-0.

Chabadel, A., I. Banon-Rodriguez, D. Cluet, B. B. Rudkin, B. Wehrle-Haller, E. Genot, P. Jurdic, I. M. Anton, and F. Saltel. 2007. "CD44 and Beta3 Integrin Organize Two Functionally Distinct Actin-Based Domains in Osteoclasts." *Molecular Biology of the Cell* 18 (12): 4899–4910. https://doi.org/10.1091/mbc.e07-04-0378.

Chang, Chong-keng, Wen-chun Hung, and Hui-chiu Chang. 2008. "The Kazal Motifs of RECK Protein Inhibit MMP-9 Secretion and Activity and Reduce Metastasis of Lung Cancer Cells *in Vitro* and *in Vivo* Gelatin Zymography" *Journal of Cellular and Molecular Medicine*, 12 (6b): 2781–89. https://doi.org/10.1111/j.1582-4934.2008.00215.x.

Changlian, Lu, Li Xiao-Yan, Hu Yuexian, Rowe R Grant, and Weiss Stephen J. 2010. "MT1-MMP Controls Human Mesenchymal Stem Cell Trafficking and Differentiation." *Blood* 115 (2): 221–29.

Cheng, Yu-Ching, Wen-Hong L. Kao, Braxton D. Mitchell, Jeffrey R. O'Connell, Haiqing Shen, Patrick F. McArdle, Quince Gibson, Kathleen A. Ryan, Alan R. Shuldiner, and Toni I. Pollin. 2009. "Genome-Wide Association Scan Identifies Variants near Matrix Metalloproteinase (MMP) Genes on Chromosome 11q21–22 Strongly Associated With Serum MMP-1 Levels." *Circulation: Cardiovascular Genetics* 2 (4): 329–37. https://doi.org/10.1161/CIRCGENETICS.108.834986.

Chevalier, X., T. Conrozier, M. Gehrmann, P. Claudepierre, P. Mathieu, S. Unger, and E. Vignon. 2001. "Tissue Inhibitor of Metalloprotease-1 (TIMP-1) Serum Level May Predict Progression of Hip Osteoarthritis."

Osteoarthritis and Cartilage 9 (4): 300–307. https://doi.org/10.1053/joca.2000.0389.

Chung, Gina G., Maciej P. Zerkowski, Idris Tolgay Ocal, Marisa Dolled-Filhart, Jung Y. Kang, Amanda Psyrri, Robert L. Camp, and David L. Rimm. 2004. "β-Catenin and P53 Analyses of a Breast Carcinoma Tissue Microarray." *Cancer* 100 (10): 2084–92. https://doi.org/10.1002/cncr.20232.

Civitelli, Roberto, Fernando Lecanda, Niklas R. Jørgensen, and Thomas H. Steinberg. 2007. "Intercellular Junctions and Cell-Cell Communication in Bone." In *Principles of Bone Biology*, 287–302. Elsevier. https://doi.org/10.1016/b978-012098652-1/50120-7.

Clarke, Bart. 2008. "Normal Bone Anatomy and Physiology." *Clinical Journal of the American Society of Nephrology : CJASN* 3 Suppl 3 (Supplement 3): S131-9. https://doi.org/10.2215/CJN.04151206.

Clokie, Cameron M, and Hershey Warshawsky. 1995. "Morphologic and Radioautographic Studies of Bone Formation in Relation to Titanium Implants Using the Rat Tibia as a Model." *The International Journal of Oral & Maxillofacial Implants* 10 (2): 155–65. http://www.ncbi.nlm.nih.gov/pubmed/7744434.

Coe, Diane Mary, Anthony William James Cooper, Paul Martin Gore, David House, Stefan Senger, and Sadie Vile. 2012 Pyrazole and triazole carboxamides as crac chann el inhibitors. WO2010122088A1. *Google Patents*, issued 2010.

Constantin, Arnaud, Valérie Lauwers-Cancès, Frédérique Navaux, Michel Abbal, Joost van Meerwijk, Bernard Mazières, Anne Cambon-Thomsen, and Alain Cantagrel. 2002. "Stromelysin 1 (Matrix Metalloproteinase 3) and HLA-DRB1 Gene Polymorphisms: Association with Severity and Progression of Rheumatoid Arthritis in a Prospective Study." *Arthritis & Rheumatism* 46 (7): 1754–62. https://doi.org/10.1002/art.10336.

Crimmin, M., A. H. Drummond, R. Gilbert, K. Miller, P. Nayee, K. Owen, W Thomas, G Wells, L M Wood, and K Woolleyt. 1995. "Matrix Metalloproteinases and Processing of Pro-TNF-α." *Journal of Leukocyte Biology* 57 (May): 774–77.

Curran, S., and G. I. Murray. 2000. "Matrix Metalloproteinases: Molecular Aspects of Their Roles in Tumour Invasion and Metastasis." *European Journal of Cancer* 36 (13): 1621–30. https://doi.org/10.1016/S0959-8049(00)00156-8.

D'Alonzo, Richard C., Nagarajan Selvamurugan, Gerard Karsenty, and Nicola C. Partridge. 2002. "Physical Interaction of the Activator Protein-1 Factors c-Fos and c-Jun with Cbfa1 for Collagenase-3 Promoter Activation." *Journal of Biological Chemistry* 277 (1): 816–22. https://doi.org/10.1074/jbc.M107082200.

Dahan, Maurice, Beatrice Nawrocki, Rene Elkaim, Martine Soell, Anne-Laure Bolcato-Bellemin, Philippe Birembaut, and Henri Tenenbaum. 2001. "Expression of Matrix Metalloproteinases in Healthy and Diseased Human Gingiva." *Journal of Clinical Periodontology* 28 (2): 128–36. https://doi.org/10.1034/j.1600-051x.2001.028002128.x.

Dallas, Sarah L., Matthew Prideaux, and Lynda F. Bonewald. 2013. "The Osteocyte: An Endocrine Cell … and More." *Endocrine Reviews* 34 (5): 658–90.

Dean, David D., Johanne Martel-Pelletier, Jean-Pierre Pelletier, David S. Howell, and J. Frederick Woessner Jr. 1989. "Evidence for Metalloproteinase Inhibitor Imbalance in Human Osteoarthritic Cartilage." *Journal of Clinical Investigation* 84 (2): 678–85. https://doi.org/10.1172/JCI114215.

Delaissé, Jean-Marie, Michael T.Engsig, Vincent Everts, Maria del Carmen Ovejero, Mercedes Ferreras, Leif Lund, Thiennu H. Vu, Zena Werb, Bent Winding, André Lochter, Morten A. Karsdal, Tine Troen, Tove Kirkegaard, Thomas Lenhard, Anne-Marie Heegaard, Lynn Neff, Roland Baron, and Niels T. Foged. 2000. "Proteinases in Bone Resorption: Obvious and Less Obvious Roles." *Clinica Chimica Acta; International Journal of Clinical Chemistry* 291 (2): 223–34. https://doi.org/10.1016/S0009-8981(99)00230-2

Delaisse, Jean-Marie, Thomas L. Andersen, Michael T. Engsig, Kim Henriksen, Tine Troen, and Laurence Blavier. 2003. "Matrix Metalloproteinases (MMP) and Cathepsin K Contribute Differently to

Osteoclastic Activities." *Microscopy Research and Technique* 61 (6): 504–13. https://doi.org/10.1002/jemt.10374.

Deleon, Kristine Y., Andriy Yabluchanskiy, Michael D. Winniford, Richard A. Lange, Robert J. Chilton, and Merry L. Lindsey. 2014. "Modifying Matrix Remodeling to Prevent Heart Failure." In *Cardiac Regeneration and Repair*, 1:41–60. Woodhead Publishing. https://doi.org/10.1533/9780857096708.1.41.

DenBesten, P. K., Y. Yan, J. D. B. Featherstone, J. F. Hilton, C. E. Smith, and W Li. 2002. "Effects of Fluoride on Rat Dental Enamel Matrix Proteinases." *Archives of Oral Biology* 47 (11): 763–70. http://www.ncbi.nlm.nih.gov/pubmed/12446183.

Dew, Gary, Gillian Murphy, Heather Stanton, Rüdiger Vallon, Peter Angel, John J. Reynolds, and Rosalind M. Hembry. 2000. "Localisation of Matrix Metalloproteinases and TIMP-2 in Resorbing Mouse Bone." *Cell and Tissue Research* 299 (3): 385–94. https://doi.org/10.1007/s004410050036.

Ducy, Patricia, and Gerard Karsenty. 1999. "Transcriptional Control of Osteoblast Differentiation." In *The Endocrinologist*, 9:32–35. Elsevier. https://doi.org/10.1097/00019616-199901000-00007.

Ducy, Patricia, Rui Zhang, Valérie Geoffroy, Amy L. Ridall, and Gérard Karsenty. 1997. "Osf2/Cbfa1: A Transcriptional Activator of Osteoblast Differentiation." *Cell* 89 (5): 747–54. https://doi.org/10.1016/S0092-8674(00)80257-3.

Eiró, Noemi, Belen Fernandez-Garcia, Luis O González, and Francisco J Vizoso. 2013. "Cytokines Related to MMP-11 Expression by Inflammatory Cells and Breast Cancer Metastasis." *Oncoimmunology* 2 (5): e24010. https://doi.org/10.4161/onci.24010.

Elmali, Nurzat, Muharrem İnan, Kadir Ertem, İrfan Esenkaya, İrfan Ayan, and Mustafa Karakaplan. 2002. "Diz Osteoartritinin Artroskopik Debridman ve Intraartikuler Hyaluronik Asit Ile Tedavisi." *Artroplasti Artroskopik Cerrahi Dergisi* 13 (3): 131–35. [Arthroscopic Debridement and Intraarticular Hyaluronic Acid Treatment of Knee Osteoarthritis. *Arthroplasty Journal of Arthroscopic Surgery* 13 (3): 131–35].

Erkli, Hande, and Engin Ersöz. 2011. "Matrix Metalloprotinases: Effects on Dental Tissues and Caries." *Cumhuriyet Dental Journal* 14 (3): 246–57. https://dergipark.org.tr/cumudj/issue/4241/56622.

Fakhry, Maya. 2013. "Molecular Mechanisms of Mesenchymal Stem Cell Differentiation towards Osteoblasts." *World Journal of Stem Cells* 5 (4): 136. https://doi.org/10.4252/wjsc.v5.i4.136.

Fejerskov, O., A. Thylsrup, and M. Joost Larsen. 1977. "Clinical and Structural Features and Possible Pathogenic Mechanisms of Dental Fluorosis." *European Journal of Oral Sciences* 85 (7): 510–34. https://doi.org/10.1111/j.1600-0722.1977.tb02110.x.

Feng, Ping, Hong Zhang, Zhuqiu Zhang, Xiaoli Dai, Ting Mao, Yinyin Fan, Xiaofang Xie, Huiyan Wen, Peijuan Yu, Yae Hu, and Ruhong Yan. 2016. "The Interaction of MMP-2/B7-H3 in Human Osteoporosis." *Clinical Immunology* 162: 118–24. https://doi.org/10.1016/j.clim.2015.11.009.

Fields, Gregg B. 1991. "A Model for Interstitial Collagen Catabolism by Mammalian Collagenases." *Journal of Theoretical Biology* 153 (4): 585–602.

Filgueira, Luis. 2010. "Osteoclast Differentiation and Function." In *Bone Cancer*, 59–66. Elsevier.

Florencio-Silva, Rinaldo, Gisela Rodrigues da Silva Sasso, Estela Sasso-Cerri, Manuel Jesus Simões, and Paulo Sérgio Cerri. 2015. "Biology of Bone Tissue: Structure, Function, and Factors That Influence Bone Cells." *BioMed Research International* 2015 (July): 421746. https://doi.org/10.1155/2015/421746.

Folgueras, Alicia R., Teresa Valdés-Sánchez, Elena Llano, Luis Menéndez, Ana Baamonde, Bristol L. Denlinger, Carlos Belmonte, Lucía Juárez, Ana Lastra, Olivia García-Suárez, Aurora Astudillo, Martina Kirstein, Alberto M. Pendás, Isabel Fariñas, and Carlos López-Otín. 2009. "Metalloproteinase MT5-MMP Is an Essential Modulator of Neuro-Immune Interactions in Thermal Pain Stimulation." *Proceedings of the National Academy of Sciences* 106 (38): 16451–56. https://doi.org/10.1073/pnas.0908507106.

Fu, Jing, Shirong Li, Rentian Feng, Huihui Ma, Farideh Sabeh, G David Roodman, Ji Wang, Samuel Robinson, X. Edward Guo, Thomas Lund, Daniel Normolle, Markus Y. Mapara, Stephen J. Weiss, and Suzanne Lentzsch. 2016. "Multiple Myeloma-Derived MMP-13 Mediates Osteoclast Fusogenesis and Osteolytic Disease." *The Journal of Clinical Investigation* 126 (5): 1759–72. https://doi.org/10.1172/JCI80276.

Galis, Zorina S, and Jaikirshan J Khatri. 2002. "Matrix Metalloproteinases in Vascular Remodeling and Atherogenesis: The Good, the Bad, and the Ugly." *Circulation Research* 90 (3): 251–62. http://www.ncbi.nlm.nih.gov/pubmed/11861412.

Gasson, Judith C., David W. Golde, Susan E. Kaufman, Carol A. Westbrook, Rodney M. Hewick, Randal J. Kaufman, Gordon G. Wong, Patricia A. Temple, Ann C. Leary, Eugene L. Brown, Elizabeth C. Orr and Steven C. Clark. 1985. "Molecular Characterization and Expression of the Gene Encoding Human Erythroid-Potentiating Activity." *Nature*, 315 (6022): 768.

Geng, Rui, Yuansheng Xu, Wenhao Hu, and Hui Zhao. 2018. "The Association between MMP-1 Gene Rs1799750 Polymorphism and Knee Osteoarthritis Risk." *Bioscience Reports* 38 (5): BSR20181257. https://doi.org/10.1042/BSR20181257.

Gentili, Chiara, and Ranieri Cancedda. 2009. "Cartilage and Bone Extracellular Matrix." *Current Pharmaceutical Design* 15 (12): 1334–48. https://doi.org/10.2174/138161209787846739.

Geoffroy, Valérie, Caroline Marty-Morieux, Nathalie Le Goupil, Phillippe Clement-Lacroix, Catherine Terraz, Monique Frain, Sophie Roux, Jérome Rossert, and Marie Christine De Vernejoul. 2004. "*In Vivo* Inhibition of Osteoblastic Metalloproteinases Leads to Increased Trabecular Bone Mass." *Journal of Bone and Mineral Research* 19 (5): 811–22. https://doi.org/10.1359/JBMR.040119.

Ghale-Noie, Zari Naderi, Mohammad Hassani, Amir R. Kachooei, and Mohammad A. Kerachian (2018). "High serum alpha-2-macroglobulin level in patients with osteonecrosis of the femoral head". *Archives of Bone and Joint Surgery*, 6 (3): 219-24

Gialeli, Chrisostomi, Achilleas D Theocharis, and Nikos K Karamanos. 2011. "Roles of Matrix Metalloproteinases in Cancer Progression and Their Pharmacological Targeting" *The FEBS journal*, 278 (1): 16–27. https://doi.org/10.1111/j.1742-4658.2010.07919.x.

Glimcher, Melvin J. 1998. "The Nature of the Mineral Phase in Bone: Biological and Clinical Implications." In *Metabolic Bone Disease and Clinically Related Disorders*, 23-52e. Elsevier. https://doi.org/10.1016/b978-012068700-8/50003-7.

Golub, L. M., M. E. Ryan, and R. C. Williams. 1998. "Modulation of the Host Response in the Treatment of Periodontitis." *Dentistry Today* 17 (10): 102–6, 108–9. http://www.ncbi.nlm.nih.gov/pubmed/10752438.

Golub, L. M., M. Wolff, S Roberts, H M Lee, M Leung, and G S Payonk. 1994. "Treating Periodontal Diseases by Blocking Tissue-Destructive Enzymes." *Journal of the American Dental Association (1939)* 125 (2): 163–69; discussion 169-71. https://doi.org/10.14219/JADA.ARCHIVE.1994.0261.

Gomis-R¨th, Franz-Xaver, Klaus Maskos, Michael Betz, Andreas Bergner, Robert Huber, Ko Suzuki, Naoki Yoshida, Hideaki Nagase, Keith Brew, Gleb P. Bourenkov, Hans Bartunik and Wolfram Bode. 1997. "Mechanism of Inhibition of the Human Matrix Metalloproteinase Stromelysin-1 by TIMP-1." *Nature* 389 (6646): 77–81. https://doi.org/10.1038/37995.

Graves, Austin R., Patricia K. Curran, Carolyn L. Smith, and Joseph A. Mindell. 2008. "The Cl-/H+ Antiporter ClC-7 Is the Primary Chloride Permeation Pathway in Lysosomes." *Nature* 453 (7196): 788–92. https://doi.org/10.1038/nature06907.

Green, J., S. Schotland, D. J. Stauber, C. R. Kleeman, and T. L. Clemens. 2017. "Cell-Matrix Interaction in Bone: Type I Collagen Modulates Signal Transduction in Osteoblast-like Cells." *American Journal of Physiology-Cell Physiology* 268 (5): C1090–1103. https://doi.org/10.1152/ajpcell.1995.268.5.c1090.

Gross, Jerome, and Charles M. Lapiere. 1962. "Collagenolytic Activity in Amphibian Tissues: A Tissue Culture Assay." *Proceedings of the*

National Academy of Sciences of the United States of America 48 (6): 1014–22. https://doi.org/10.1073/pnas.48.6.1014.

Grossman, Moran, Dmitry Tworowski, Orly Dym, Meng Huee Lee, Yaakov Levy, Gillian Murphy, and Irit Sagi. 2010. "The Intrinsic Protein Flexibility of Endogenous Protease Inhibitor TIMP-1 Controls Its Binding Interface and Affects Its Function." *Biochemistry* 49 (29): 6184–92. https://doi.org/10.1021/bi902141x.

Guo, Feng, Junsheng Feng, Feng Wang, Wentong Li, Qingping Gao, Zhuo Chen, Lisa Shoff, Kevin J. Donly, Jelica Gluhak-Heinrich, Yong Hee Patricia Chun, Stephen E. Harris, Mary MacDougall, and Shuo Chen. 2015. "Bmp2 Deletion Causes an Amelogenesis Imperfecta Phenotype Via Regulating Enamel Gene Expression." *Journal of Cellular Physiology* 230 (8): 1871–82. https://doi.org/10.1002/jcp.24915.

Haeusler, G., I. Walter, M. Helmreich, and M. Egerbacher. 2005. "Localization of Matrix Metalloproteinases, (MMPs) Their Tissue Inhibitors, and Vascular Endothelial Growth Factor (VEGF) in Growth Plates of Children and Adolescents Indicates a Role for MMPs in Human Postnatal Growth and Skeletal Maturation." *Calcified Tissue International* 76 (5): 326–35. https://doi.org/10.1007/s00223-004-0161-6.

Hanemaaijer, Roeland, Timo Sorsa, Yrjö T. Konttinen, Yanli Ding, Meeri Sutinen, Hetty Visser, Victor W. M. van Hinsbergh, Tarja Helaakoski, Tiina Kainulainen, Hanne Rönkä, Harald Tschesche and Tuula Salo. 1997. "Matrix Metalloproteinase-8 Is Expressed in Rheumatoid Synovial Fibroblasts and Endothelial Cells Regulation by Tumor Necrosis Factor-Alpha and Doxycycline." *The Journal of Biological Chemistry* 272 (50): 31504–9. https://doi.org/10.1074/jbc.272.50.31504.

Hannas, Angélica R., José C. Pereira, José M. Granjeiro, and Leo Tjäderhane. 2007. "The Role of Matrix Metalloproteinases in the Oral Environment." *Acta Odontologica Scandinavica* 65 (1): 1–13. https://doi.org/10.1080/00016350600963640.

Haq, I., E. Murphy, and J. Dacre. 2003. "Osteoarthritis." *Postgraduate Medical Journal* 79 (933): 377–83. https://doi.org/10.1136/pmj.79.933.377.

Hikita, Atsuhiko, Ikuo Yana, Hidetoshi Wakeyama, Masaki Nakamura, Yuho Kadono, Yasushi Oshima, Kozo Nakamura, Motoharu Seiki, and Sakae Tanaka. 2006. "Negative Regulation of Osteoclastogenesis by Ectodomain Shedding of Receptor Activator of NF-KB Ligand." *Journal of Biological Chemistry* 281 (48): 36846–55. https://doi.org/10.1074/jbc.M606656200.

Holmbeck, Kenn, Paolo Bianco, John Caterina, Susan Yamada, Mark Kromer, Sergei A. Kuznetsov, Mahesh Mankani, Pamela Gehron Robey, A. Robin Poole, Isabelle Pidoux, Jerrold M. Ward, and Henning Birkedal-Hansen. 1999. "MT1-MMP-Deficient Mice Develop Dwarfism, Osteopenia, Arthritis, and Connective Tissue Disease Due to Inadequate Collagen Turnover." *Cell* 99 (1): 81-92. https://doi.org/10.1016/S0092-8674(00)80064-1.

Hubmacher, Dirk, and Suneel S. Apte. 2013. "The Biology of the Extracellular Matrix: Novel Insights." *Current Opinion in Rheumatology* 25 (1): 65–70. https://doi.org/10.1097/BOR.0b013e32835b137b.

Hunziker, Ernst B. 1994. "Mechanism of Longitudinal Bone Growth and Its Regulation by Growth Plate Chondrocytes." *Microscopy Research and Technique* 28 (6): 505–19. https://doi.org/10.1002/jemt.1070280606.

Illman, Sara A., Jouko Lohi, and Jorma Keski-Oja. 2008. "Epilysin (MMP-28)-Structure, Expression and Potential Functions." *Experimental Dermatology* 17 (11): 897–907. https://doi.org/10.1111/j.1600-0625.2008.00782.x.

Imai, K, Y Yokohama, I Nakanishi, E Ohuchi, Y Fujii, N Nakai, and Y Okada. 1995. "Matrix Metalloproteinase 7 (Matrilysin) from Human Rectal Carcinoma Cells activation of the precursor, interaction with other matrix metalloproteinases and enzymic properties." *Journal of Biological Chemistry* 270 (12): 6691-6697. https://doi.org/10.1074/jbc.270.12.6691

Inada, M, M Byrne, M Yu, K Hirose, C Miyaura, C Lopezotin, S Shapiro, and S Krane. 2006. "Collagenases in Skeletal Development and Remodeling." *Matrix Biology* 25 (25): S49–50. https://doi.org/10.1016/j.matbio.2006.08.137.

Ingman, Tuula, Satu Apajalahti, Päivi Mäntylä, Pirjo Savolainen, and Timo Sorsa. 2005. "Matrix Metalloproteinase-1 and -8 in Gingival Crevicular Fluid during Orthodontic Tooth Movement: A Pilot Study during 1 Month of Follow-up after Fixed Appliance Activation." *European Journal of Orthodontics* 27 (2): 202–7. https://doi.org/10.1093/ejo/cjh097.

Inoue, Keiichi, Yuko Mikuni-Takagaki, Kaoru Oikawa, Takeshi Itoh, Masaki Inada, Takanori Noguchi, Jin-Sung Park, Takashi Onodera, Stephen M. Krane, Masaki Noda, and Shigeyoshi Itohara 2006. "A Crucial Role for Matrix Metalloproteinase 2 in Osteocytic Canalicular Formation and Bone Metabolism." *Journal of Biological Chemistry* 281 (44): 33814–24. https://doi.org/10.1074/jbc.M607290200.

Ishikawaa, Takashi, Yasushi Ichikawa, Masato Mitsuhashi, Nobuyoshi Momiyama, Takashi Chishima, Kuniya Tanaka, Hiroyuki Yamaoka, Kaoru Miyazaki, Yoji, Nagashima, Tatsuo Akitaya, and Hiroshi Shimada. 1996. "Matrilysin Is Associated with Progression of Colorectal Tumor." *Cancer Letters* 107 (1): 5–10. https://doi.org/10.1016/0304-3835(96)04336-4.

Itoh, Yoshifumi. 2017. "Metalloproteinases in Rheumatoid Arthritis: Potential Therapeutic Targets to Improve Current Therapies." In *Progress in Molecular Biology and Translational Science*. (Vol. 148, pp. 327-338). Academic Press. https://doi.org/10.1016/bs.pmbts.2017.03.002.

Jeon, Hye Won, Kyung-ju Lee, Sun Hee Lee, Woo-ho Kim, and You Mie Lee. 2011. "Attenuated Expression and Function of the RECK Tumor Suppressor Under Hypoxic Conditions Is Mediated by the MAPK Signaling Path- Ways" *Archives of pharmacal research* 34 (1): 137–45. https://doi.org/10.1007/s12272-011-0116-1.

Jia, Feng, Yu Hua Yin, Guo Yi Gao, Yu Wang, Lian Cen, and Ji-yao Jiang. 2014. "MMP-9 Inhibitor SB-3CT Attenuates Behavioral Impairments and Hippocampal Loss after Traumatic Brain Injury in Rat." *Journal of Neurotrauma* 31 (13): 1225–34. https://doi.org/10.1089/neu.2013.3230.

Jilka, Robert L., Robert S. Weinstein, Teresita Bellido, A. Michael Parfitt, and Stavros C Manolagas. 1998. "Osteoblast Programmed Cell Death (Apoptosis): modulation by growth factors and cytokines." *Journal of Bone and Mineral Research* 13 (5): 793–802.

Jiménez, M. J., M Balbín, J. M. López, J Alvarez, T. Komori, and C. López-Otín. 1999. "Collagenase 3 Is a Target of Cbfa1, a Transcription Factor of the Runt Gene Family Involved in Bone Formation." *Molecular and Cellular Biology* 19 (6): 4431–42.

Jones, C., David C. Sane, and David M. Herrington. 2003. "Matrix Metalloproteinases A Review of Their Structure and Role in Acute Coronary Syndrome." *Cardiovascular Research* 59 (4): 812–23. https://doi.org/10.1016/S0008-6363(03)00516-9.

Kaneshiro, Bliss, Alison Edelman, Chandravanu Dash, Jui Pandhare, Faapisa M. Soli, and Jeffrey T. Jensen. 2016. "Effect of Oral Contraceptives and Doxycycline on Endometrial MMP-2 and MMP-9 Activity." *Contraception* 93 (1): 65–69. https://doi.org/10.1016/j.contraception.2015.09.006.

Kara, Murat, Ahmet Imerci, Umut Canbek, Ulas Akgun, Tugba Dubektas Canbek, and Nevres Hurriyet Aydogan. 2016. "The Relationship Between Primary Knee Osteoarthritis (OA) and MMP-1 and MMP-3 Gene Polymorphisms in a Turkish Population: A Population-Based Case-Control Study." *Muğla Sıtkı Koçman Üniversitesi Tıp Dergisi/Mugla Medical Journal* 3 (2): 9–12.

Karsdal, M. A., T. A. Andersen, L. Bonewald, and C. Christiansen. 2004. "Matrix Metalloproteinases (MMPs) Safeguard Osteoblasts from Apoptosis during Transdifferentiation into Osteocytes: MT1-MMP Maintains Osteocyte Viability." *DNA and Cell Biology* 23 (3): 155–65. https://doi.org/10.1089/104454904322964751.

Kaspiris, A., L. Khaldi, T. B. Grivas, E. Vasiliadis, I. Kouvaras, S. Dagkas, E. Chronopoulos, and E. Papadimitriou. 2013. "Subchondral Cyst Development and MMP-1 Expression during Progression of Osteoarthritis: An Immunohistochemical Study." *Orthopaedics &*

Traumatology: Surgery & Research 99 (5): 523–29. https://doi.org/10.1016/J.OTSR.2013.03.019.

Kenkre, Julia S., and John H. D. Bassett. 2018. "The Bone Remodelling Cycle." *Annals of Clinical Biochemistry* 55 (3): 308–27. https://doi.org/10.1177/0004563218759371.

Kessenbrock, Kai, Vicki Plaks, and Zena Werb. 2010. "Matrix Metalloproteinases: Regulators of the Tumor Microenvironment." *Cell.* 141 (1): 52-67 https://doi.org/10.1016/j.cell.2010.03.015.

Kim, J. W., J. P. Simmer, T. C. Hart, P. S. Hart, M. D. Ramaswami, J. D. Bartlett, and J. C. C. Hu. 2005. "MMP-20 Mutation in Autosomal Recessive Pigmented Hypomaturation Amelogenesis Imperfecta." *Journal of Medical Genetics* 42 (3): 271–75. https://doi.org/10.1136/jmg. 2004.024505.

Kim, Youhwan, Hyojin Ko, Ik Keun Kwon, and Kwanwoo Shin. 2016. "Extracellular Matrix Revisited: Roles in Tissue Engineering." *International Neurourology Journal* 20: S23–29. https://doi.org/10.5213/inj.1632600.318.

Knox, J. D., D. R. Boreham, J. A. Walker, D. P. Morrison, L. M. Matrisian, R. B. Nagle, and G. T. Bowden. 1996. "Mapping of the Metalloproteinase Gene Matrilysin (MMP7) to Human Chromosome 11q21↠Q22." *Cytogenetic and Genome Research* 72 (2–3): 179–82. https://doi.org/10.1159/000134181.

Komori, T., H. Yagi, S. Nomura, A. Yamaguchi, K. Sasaki, K. Deguchi, Y. Shimizu, R. T. Bronson, Y.-H. Gao, M. Inada, M. Sato, R. Okamot, Y. Kitamura, S. Yoshiki, and T. Kishimoto. 1997. "Targeted Disruption of Cbfa1 Results in a Complete Lack of Bone Formation Owing to Maturational Arrest of Osteoblasts." *Cell* 89 (5): 755–64. https://doi.org/10.1016/S0092-8674(00)80258-5.

Konttinen, Yrjö T., Mia Ainola, Heikki Valleala, Jia Ma, Hideo Ida, Jami Mandelin, Raimund W. Kinne, Seppo Santavirtae, Timo Sorsaf, Carlos López-Otíng, and Michiaki Takagic. 1999. "Analysis of 16 Different Matrix Metalloproteinases (MMP-1 to MMP-20) in the Synovial Membrane: Different Profiles in Trauma and Rheumatoid Arthritis."

Annals of the Rheumatic Diseases 58 (11): 691–97. https://doi.org/10.1136/ard.58.11.691.

Kraft-Neumärker, M., K. Lorenz, R. Koch, T. Hoffmann, P. Mäntylä, T. Sorsa, and L. Netuschil. 2012. "Full-Mouth Profile of Active MMP-8 in Periodontitis Patients." *Journal of Periodontal Research* 47 (1): 121–28. https://doi.org/10.1111/j.1600-0765.2011.01416.x.

Kramer, I., C. Halleux, H. Keller, M. Pegurri, J. H. Gooi, P. B. Weber, J. Q. Feng, L. F. Bonewald, and M. Kneissel. 2010. "Osteocyte Wnt/-Catenin Signaling Is Required for Normal Bone Homeostasis." *Molecular and Cellular Biology* 30 (12): 3071–85. https://doi.org/10.1128/MCB.01428-09.

Krane, Stephen, and Masaki Inada. 2008. "Matrix Metalloproteinases and Bone." *Bone* 43 (1): 7–18. https://doi.org/10.1016/j.bone.2008.03.020.

Krane, Stephen M. 2001. "Petulant Cellular Acts: Destroying the ECM Rather than Creating It." *The Journal of Clinical Investigation* 107 (1): 31–32. https://doi.org/10.1172/JCI11892.

Kurzepa, Jacek, Marcin Baran, Slawomir Watroba, Malgorzata Barud, and Daniel Babula. 2014. "Collagenases and Gelatinases in Bone Healing. The Focus on Mandibular Fractures." *Current Issues in Pharmacy and Medical Sciences* 27 (2): 121–26. https://doi.org/10.2478/cipms-2014-0029.

Kusano, Kenichiro, Chisato Miyaura, Masaki Inada, Tatsuya Tamura, Akira Ito, Hideaki Nagase, Kyuichi Kamoi, and Tatsuo Suda. 1998. "Regulation of Matrix Metalloproteinases (MMP-2,-3,-9, and -13) by Interleukin-1 and Interleukin-6 in Mouse Calvaria: Association of MMP Induction with Bone Resorption." *Endocrinology* 139 (3): 1338–45. https://doi.org/10.1210/endo.139.3.5818.

Lee, W., S. Aitken, J. Sodek, and C. A. G. McCulloch. 1995. "Evidence of a Direct Relationship between Neutrophil Collagenase Activity and Periodontal Tissue Destruction *in Vivo*: Role of Active Enzyme in Human Periodontitis." *Journal of Periodontal Research* 30 (1): 23–33. https://doi.org/10.1111/j.1600-0765.1995.tb01249.x.

Levy, Annie, Jessica Zucman, Olivier Delattre, Marie-Genevie`ve Mattei, Marie-Christine Rio, and Paul Basset. 1992. "Assignment of the Human

Stromelysin 3 (STMY3) Gene to the Q11.2 Region of Chromosome 22." *Genomics* 13 (3): 881–83. https://doi.org/10.1016/0888-7543(92)90175-R.

Li, Heng, Dan Wang, Yongjian Yuan, and Jikang Min. 2017. "New Insights on the MMP-13 Regulatory Network in the Pathogenesis of Early Osteoarthritis." *Arthritis Research and Therapy* 19 (1): 1–12. https://doi.org/10.1186/s13075-017-1454-2.

Li, N. G., Z. H. Shi, Y. P. Tang, Z. J. Wang, S. L. Song, L. H. Qian, D. W. Qian, and J. A. Duan. 2011. "New Hope for the Treatment of Osteoarthritis Through Selective Inhibition of MMP-13." *Current Medicinal Chemistry* 18 (7): 977–1001. https://doi.org/10.2174/092986711794940905.

Liang, Hai Po Helena, Joshua Xu, Meilang Xue, and Christopher Jackson. 2016. "Matrix Metalloproteinases in Bone Development and Pathology: Current Knowledge and Potential Clinical Utility." *Metalloproteinases In Medicine* 3 (December): 93–102. https://doi.org/10.2147/MNM.S92187.

Liang, Tangzhao, Wenling Gao, Lei Zhu, Jianhua Ren, Hui Yao, Kun Wang, and Dehai Shi. 2019. "TIMP-1 Inhibits Proliferation and Osteogenic Differentiation of HBMSCs through Wnt/β-Catenin Signaling." *Bioscience Reports* 39 (1): BSR20181290. https://doi.org/10.1042/bsr20181290.

Liao, Er-yuan, and Xiang-hang Luo. 2001. "Effects of 17β-Estradiol on the Expression of Matrix Metalloproteinase-1, -2 and Tissue Inhibitor of Metalloproteinase-1 in Human Osteoblast-like Cells Cultures." *Endocrine* 15 (3): 291–96. https://doi.org/10.1385/ENDO:15:3:291.

Lint, Philippe Van, and Claude Libert. 2006. "Matrix Metalloproteinase-8: Cleavage Can Be Decisive." *Cytokine & Growth Factor Reviews* 17 (4): 217–23. https://doi.org/10.1016/J.CYTOGFR.2006.04.001.

Liu, T. W., M. K. Akens, J. Chen, B. C. Wilson, and G. Zheng. 2016. "Matrix Metalloproteinase-Based Photodynamic Molecular Beacons for Targeted Destruction of Bone Metastases *in Vivo*." *Photochemical and Photobiological Sciences* 15 (3): 375–81. https://doi.org/10.1039/c5pp00414d.

Löffek, Stefanie, Oliver Schilling, and Claus-Werner Franzke. 2011. "Biological Role of Matrix Metalloproteinases: A Critical Balance." *European Respiratory Journal* 38 (1): 191–208. https://doi.org/10.1183/09031936.00146510.

Loreto, C, R Leonardi, G Musumeci, G Pannone, and S Castorina. 2013. "An Ex Vivo Study on Immunohistochemical Localization of MMP-7 and MMP-9 in Temporomandibular Joint Discs with Internal Derangement." *European Journal of Histochemistry : EJH* 57 (2): e12. https://doi.org/10.4081/ejh.2013.e12.

Lories, Rik J. U., Jenny Peeters, Astrid Bakker, Przemko Tylzanowski, Inge Derese, Jan Schrooten, J. Terrig Thomas, and Frank P. Luyten. 2007. "Articular Cartilage and Biomechanical Properties of the Long Bones in *Frzb* -Knockout Mice." *Arthritis & Rheumatism* 56 (12): 4095–4103. https://doi.org/10.1002/art.23137.

Lozito, Thomas P., Wesley M. Jackson, Leon J. Nesti, and Rocky S. Tuan. 2014. "Human Mesenchymal Stem Cells Generate a Distinct Pericellular Zone of MMP Activities via Binding of MMPs and Secretion of High Levels of TIMPs." *Matrix Biology*. 34: 132-143. https://doi.org/10.1016/j.matbio.2013.10.003.

Lozito, Thomas P., and Rocky S. Tuan. 2011. "Mesenchymal Stem Cells Inhibit Both Endogenous and Exogenous MMPs via Secreted TIMPs." *Journal of Cellular Physiology*. 226 (2): 385-396. https://doi.org/10.1002/ jcp.22344.

Luo, Shufang, Mohong Deng, Xing Long, Jian Li, Liqin Xu, and Wei Fang. 2015. "Association between Polymorphism of MMP-1 Promoter and the Susceptibility to Anterior Disc Displacement and Temporomandibular Joint Osteoarthritis." *Archives of Oral Biology* 60 (11): 1675–80. https://doi.org/10.1016/j.archoralbio.2015.08.001.

Luxenburg, Chen, Dafna Geblinger, Eugenia Klein, Karen Anderson, Dorit Hanein, Benny Geiger, and Lia Addadi. 2007. "The Architecture of the Adhesive Apparatus of Cultured Osteoclasts: From Podosome Formation to Sealing Zone Assembly." *PLoS ONE* 2 (1): e179. https://doi.org/10.1371/journal.pone.0000179.

Lynch, Conor C., Atsuya Hikosaka, Heath B. Acuff, Michelle D. Martin, Noriyasu Kawai, Rakesh K. Singh, Tracy C. Vargo-Gogola, Jennifer L. Begtrup, Todd E. Peterson, Barbara Fingleton, Tomoyuki Shirai, Lynn M. Matrisian, and Mitsuru Futakuchi. 2005. "MMP-7 Promotes Prostate Cancer-Induced Osteolysis via the Solubilization of RANKL." *Cancer Cell* 7 (5): 485–96. https://doi.org/10.1016/j.ccr.2005.04.013

Malemud, Charles, J. 2006. "Matrix Metalloproteinases: Role in Skeletal Development and Growth Plate Disorders." *Frontiers in Bioscience* 11 (1): 1702. https://doi.org/10.2741/1916.

Mäntylä, Päivi, Mathias Stenman, Denis F Kinane, Sari Tikanoja, Hanne Luoto, Tuula Salo, and Timo Sorsa. 2003. "Gingival Crevicular Fluid Collagenase-2 (MMP-8) Test Stick for Chair-Side Monitoring of Periodontitis." *Journal of Periodontal Research* 38 (4): 436–39. https://doi.org/10.1034/j.1600-0765.2003.00677.x

Martignetti, John A., Aida Al Aqeel, Wafaa Al Sewairi, Christine E. Boumah, Marios Kambouris, S. Al Mayouf, K. V. Sheth, W. Al Eid, Oonagh Dowling, Juliette Harris, Marc J. Glucksman, Sultan Bahabri, Brian F. Meyer, and Robert J. Desnick. 2001. "Mutation of the Matrix Metalloproteinase 2 Gene (MMP2) Causes a Multicentric Osteolysis and Arthritis Syndrome." *Nature Genetics* 28 (3): 261–65. https://doi.org/10.1038/90100.

Maskos, Klaus, and Wolfram Bode. 2003. "Structural Basis of Matrix Metalloproteinases and Tissue Inhibitors of Metalloproteinases." *Applied Biochemistry and Biotechnology-Part B Molecular Biotechnology* 25 (3): 241–66. https://doi.org/10.1385/MB:25:3:241.

Masson,Régis, Olivier Lefebvre, Agnès Noël, Mostapha El Fahime, Marie-Pierre Chenard, Corinne Wendling, Florence Kebers, Marianne LeMeur, Andrée Dierich, Jean-Michel Foidart, Paul Basset, and Marie-Christine Rio. 1998. "*In Vivo* Evidence That the Stromelysin-3 Metalloproteinase Contributes in a Paracrine Manner to Epithelial Cell Malignancy." *The Journal of Cell Biology* 140 (6): 1535–41. https://doi.org/10.1083/jcb.140.6.1535.

Matrisian, L. M. 1990. "Metalloproteinases and Their Inhibitors in Matrix Remodeling." *Trends in Genetics : TIG* 6 (4): 121–25. https://doi.org/10.1016/0168-9525(90)90126-Q

Matrisian, Lynn M. 1992. "The Matrix-Degrading Metalloproteinases." *BioEssays* 14 (7): 455–63. https://doi.org/10.1002/bies.950140705.

Matthews, Jessica Lester, and R. V. Talmage. 1981. "Influence of Parathyroid Hormone on Bone Cell Ultrastructure." *Clinical Orthopaedics and Related Research* NA; (156): 27–38. https://doi.org/10.1097/00003086-198105000-00005.

McIntush, E. 2004. "Matrix Metalloproteinases and Tissue Inhibitors of Metalloproteinases in Ovarian Function." *Reviews of Reproduction* 3 (1): 23–30. https://doi.org/10.1530/revreprod/3.1.23.

Mecham, Robert P., and William C. Parks. 1998. *Matrix Metalloproteinases*. Elsevier.

Meikle, M. C., S. Bord, R. M. Hembry, J. Compston, P. I. Croucher, and J. J. Reynolds. 1992. "Human Osteoblasts in Culture Synthesize Collagenase and Other Matrix Metalloproteinases in Response to Osteotropic Hormones and Cytokines." *Journal of Cell Science* 103 (Pt 4) (December): 1093–99. http://www.ncbi.nlm.nih.gov/pubmed/1336777.

Mocchegiani, Eugenio, Robertina Giacconi, Elisa Muti, and Mario Muzzioli. 2004. "Zinc-Binding Proteins (Metallothionein and α-2 Macroglobulin) as Potential Biological Markers of Immunosenescence." In *NeuroImmune Biology*, 4:23–40. Elsevier Masson SAS. https://doi.org/10.1016/S1567-7443(04)80004-8.

Munhoz, F. B. A., A. L. Godoy-Santos, and M. C. Santos. 2010. "MMP-3 Polymorphism: Genetic Marker in Pathological Processes (Review)." *Molecular Medicine Reports* 3 (5): 735–40. https://doi.org/10.3892/mmr.2010.340.

Murate, T., and T. Hayakawa. 1999. "Multiple Functions of Tissue Inhibitors of Metalloproteinases (TIMPs): New Aspects in Hematopoiesis." *Platelets* 10 (1): 5–16. https://doi.org/10.1080/09537109976293.

Murphy, Gillian 2015. "Matrix Metalloproteinases." *Encyclopedia of Cell Biology* 1: 621–29. https://doi.org/10.1016/B978-0-12-394447-4.10073-2.

Nagase, Hideaki, Robert Visse, and Gillian Murphy. 2006. "Structure and Function of Matrix Metalloproteinases and TIMPs." *Cardiovascular Research* 69 (3): 562–73. https://doi.org/10.1016/j.cardiores.2005.12.002.

Nakano, Yukiko, William N. Addison, and Mari T. Kaartinen. 2007. "ATP-Mediated Mineralization of MC3T3-E1 Osteoblast Cultures." *Bone* 41 (4): 549–61. https://doi.org/10.1016/J.BONE.2007.06.011.

Nakatani, Teruyo, Tiffany Chen, and Nicola C Partridge. 2016. "MMP-13 Is One of the Critical Mediators of the Effect of HDAC4 Deletion on the Skeleton." *Bone* 90: 142–51. https://doi.org/10.1016/j.bone.2016.06.010.

Nampei, Akihide, Jun Hashimoto, Kenji Hayashida, Hideki Tsuboi, Kenrin Shi, Isamu Tsuji, Hideaki Miyashita, Takao Yamada, Naomichi Matsukawa, Masayuki Matsumoto, Shigeto Morimoto, Toshio Ogihara, Takahiro Ochi, and Hideki Yoshikawa. 2004. "Matrix Extracellular Phosphoglycoprotein (MEPE) Is Highly Expressed in Osteocytes in Human Bone." *Journal of Bone and Mineral Metabolism* 22 (3): 176–84. https://doi.org/10.1007/s00774-003-0468-9.

Nomura, T, A Ishii, Y Oishi, H Kohma, and K Hara. 2008. "Tissue Inhibitors of Metalloproteinases Level and Collagenase Activity in Gingival Crevicular Fluid: The Relevance to Periodontal Diseases." *Oral Diseases* 4 (4): 231–40. https://doi.org/10.1111/j.1601-0825.1998.tb00286.x.

Nonsrijun, Nongnuch, Jumphol Mitchai, Kamoltip Brown, Ratana Leksomboon, and Panya Tuamsuk. 2013. "Overexpression of Matrix Metalloproteinase 11 in Thai Prostatic Adenocarcinoma Is Associated with Poor Survival." *Asian Pacific Journal of Cancer Prevention : APJCP* 14 (5): 3331–35. https://doi.org/10.7314/apjcp.2013.14.5.3331.

Oh, Junseo, Rei Takahashi, Shunya Kondo, Akira Mizoguchi, Eijiro Adachi, Regina M Sasahara, Sachiko Nishimura, Yukio Imamura, Hitoshi Kitayama, David B. Alexander, Chizuka Ide, Thomas P. Horan,

Tsutomu Arakawa, Hisahito Yoshida, Shin-ichi Nishikawa, Yoshifumi Itoh, Motoharu Seiki, Shigeyoshi Itohara, Chiaki Takahashi, and Makoto, Noda. 2001. "The Membrane-Anchored MMP Inhibitor RECK Is a Key Regulator of Extracellular Matrix Integrity and Angiogenesis" *Cell* 107 (6): 789–800. https://doi.org/10.1016/S0092-8674(01)00597-9

Ohno, Koji, Swen Hu, Wencke Armsen, Volker Eulenburg, Diethelm W Richter, Bodo Laube, Heinrich Betz, and Glyt Gene. 2001. "Loss of the ClC-7 Chloride Channel Leads to Osteopetrosis in Mice and Man." *Cell* 104 (2): 205–15. https://doi.org/10.1016/S0968-0004(05)00043-5.

Okada, Yasunori. 2017. "Proteinases and Matrix Degradation." In *Kelley and Firestein's Textbook of Rheumatology*, 106–25. Elsevier. https://doi.org/10.1016/B978-0-323-31696-5.00008-5.

Öncel, Müfide. 2012. "Matriks Metalloproteinazlar ve Kanser." *European Journal of Basic. Medical Sciences*, 2 (3): 91–100. ["Matrix Metalloproteinases and Cancer." *European Journal of Basic. Medical Sciences*, 2 (3): 91–100].

Paiva, Katiucia Batista Silva, and José Mauro Granjeiro. 2017. "Matrix Metalloproteinases in Bone Resorption, Remodeling, and Repair." In *Matrix Metalloproteinases and Tissue Remodeling in Health and Disease: Target Tissues and Therapy*, 148: 203–303. Elsevier. https://doi.org/10.1016/bs.pmbts.2017.05.001.

Palosaari, Heidi, Caroline J. Pennington, Markku Larmas, Dylan R. Edwards, Leo Tjaderhane, and Tuula Salo. 2003. "Expression Profile of Matrix Metalloproteinases (MMPs) and Tissue Inhibitors of MMPs in Mature Human Odontoblasts and Pulp Tissue." *European Journal of Oral Sciences* 111 (2): 117–27. https://doi.org/10.1034/j.1600-0722.2003.00026.x.

Pap, Thomas, and Adelheid Korb-Pap. 2015. "Cartilage Damage in Osteoarthritis and Rheumatoid Arthritis-Two Unequal Siblings." *Nature Reviews Rheumatology*, 11 (10): 606-15 https://doi.org/10.1038/nrrheum.2015.95.

Parfitt, A. Michael. 1977. "The cellular basis of bone turnover and bone loss: a rebuttal of the osteocytic resorption--bone flow theory." *Clinical*

Orthopaedics and Related Research, 127: 236-47. https://doi.org/10.1097/00003086-197709000-00036.

Parfitt, A.M. 1990. "Bone-forming cells in clinical conditions." In: Hall BK (ed) *Bone. The Osteoblast and Osteocyte*. Telford Press and CRC Press, Boca Raton, FL, vol 1:351–429

Parvizi, Javad. 2010. "Osteoclasts." In *High Yield Orthopaedics E-Book: Expert Consult-Online and Print*, 337–39. Elsevier Health Sciences.

Pereira Prado, Vanesa, Natalia Asquino, Delmira Apellaniz, Luis Bueno Rossy, Gabriel Tapia, and Ronell Bologna Molina. 2016. "Metaloproteinasas de La Matriz Extracelular (MMPs) En Odontología." *Odontoestomatología* 18(28): 20–29. ["Extracellular Matrix Metalloproteinases (MMPs) in Dentistry" *Pediatric dentistry* 18 (28): 20–29].

Phan, T. C. A., J. Xu, and Ming H. Zheng. 2004. "Interaction between Osteoblast and Osteoclast: Impact in Bone Disease." *Histology and Histopathology* 19 (4): 1325–44.

Piert, Morand, Hans-Jürgen Machulla, Maria Picchio, Gerald Reischl, Sybille Ziegler, Piyush Kumar, Hans-Jürgen Wester, Roswitha Beck, Alexander J.B. McEwan, Leonard I. Wiebe, and Markus Schwaiger. 2005. "Hypoxia-Specific Tumor Imaging with F-18-Fluoroazomycin Arabinoside." *Journal of Nuclear Medicine* 46 (1): 106–13.

Pockwinse, S., D. Conlon, J. B. Lian, and G. S. Stein. 2004. "Expression of Cell Growth and Bone Specific Genes at Single Cell Resolution during Development of Bone Tissue-like Organization in Primary Cultures of Normal Diploid Osteoblasts." *Bone* 13 (5): A27–A27. https://doi.org/10.1016/8756-3282(92)90565-e.

Poulet, Blandine, Ke Liu, Darren Plumb, Phoung Vo, Mittal Shah, Katherine Staines, Alexandra Sampson, Hiroyuki Nakamura, Hideaki Nagase, Alessandra Carriero, Sandra Shefelbine, Andrew A. Pitsillides, and George Bou-Gharios. 2016. "Overexpression of TIMP-3 in Chondrocytes Produces Transient Reduction in Growth Plate Length but Permanently Reduces Adult Bone Quality and Quantity." *PLoS ONE* 11 (12): 1–19. https://doi.org/10.1371/journal.pone.0167971

Prideaux, M., K.A. A. Staines, E.R. R. Jones, G.P. P. Riley, A.A. A. Pitsillides, and C. Farquharson. 2015. "MMP and TIMP Temporal Gene Expression during Osteocytogenesis." *Gene Expression Patterns* 18 (1–2): 29–36. https://doi.org/10.1016/j.gep.2015.04.004.

Pufe, Thomas, Viola Harde, Wolf Petersen, Mary B. Goldring, Bernhard Tillmann, and Rolf Mentlein. 2004. "Vascular Endothelial Growth Factor (VEGF) Induces Matrix Metalloproteinase Expression in Immortalized Chondrocytes." *Journal of Pathology* 202 (3): 367–74. https://doi.org/10.1002/path.1527.

Pullen, Nicholas, Andrew Pickford, Mark Perry, Diane Jaworski, Katie Loveson, Daniel Arthur, Jonathan Holliday, Timothy E. Van Meter, Ritchie Peckham, Waqar Younas, Sophie E.J. Briggs, Sophie MacDonald, Thomas Butterfield, Myrianni Constantinou, and Helen L. Fillmore. 2018. "Current Insights into Matrix Metalloproteinases and Glioma Progression: Transcending the Degradation Boundary." *Metalloproteinases In Medicine* 5: 13–30. https://doi.org/10.2147/MNM.S105123.

Reel, Buket. 2006. "Matrix Metalloproteinases and Atherosclerosis: Review." *Turkiye Klinikleri Journal of Medical Sciences* 26 (5): 527–37.

Ricard-Blum, Sylvie, and Romain Salza. 2014. "Matricryptins and Matrikines: Biologically Active Fragments of the Extracellular Matrix." *Experimental Dermatology* 23 (7): 457-463. https://doi.org/10.1111/exd.12435.

Rochefort, G. Y., S. Pallu, and C. L. Benhamou. 2010. "Osteocyte: The Unrecognized Side of Bone Tissue." *Osteoporosis International* 21 (9): 1457–69. https://doi.org/10.1007/s00198-010-1194-5.

Rodan, Gideon A., and Shun Ichi Harada. 1997. "The Missing Bone." *Cell* 89 (5): 677–80. https://doi.org/10.1016/S0092-8674(00)80249-4.

Romanelli, Raquel, Sabrina Mancini, Carol Laschinger, Christopher M. Overall, Jaro Sodek, and Christopher A. G. McCulloch. 1999. "Activation of Neutrophil Collagenase in Periodontitis." *Infection and Immunity* 67 (5): 2319–26. http://www.ncbi.nlm.nih.gov/pubmed/10225890.

Roscilli, Giuseppe, Manuela Cappelletti, Claudia De Vitis, Gennaro Ciliberto, Arianna Di Napoli, Luigi Ruco, Rita Mancini, and Luigi Aurisicchio. 2014. "Circulating MMP11 and Specific Antibody Immune Response in Breast and Prostate Cancer Patients." *Journal of Translational Medicine* 12 (1): 54. https://doi.org/10.1186/1479-5876-12-54.

Rose, Brandon J., and David L. Kooyman. 2016. "A Tale of Two Joints: The Role of Matrix Metalloproteases in Cartilage Biology." *Disease Markers* 2016: 1–7. https://doi.org/10.1155/2016/4895050.

Ross, F. Patrick. 2006. "M-CSF, c-Fms, and Signaling in Osteoclasts and Their Precursors." *Annals of the New York Academy of Sciences* 1068 (1): 110–16. https://doi.org/10.1196/annals.1346.014.

Ross, F Patrick. 2011. "Osteoclasts." In *Vitamin D*, edited by David Feldman, J. Wesley Pike, and John S Adams, 335–47. Elsevier.

Safadi, Fayez F., Mary F. Barbe, Samir M. Abdelmagid, Mario C. Rico, Rulla A. Aswad, Judith Litvin, and Steven N. Popoff. 2009. "Bone Structure, Development and Bone Biology." In *Bone Pathology*, 1–50. Springer. https://doi.org/10.1007/978-1-59745-347-9_1.

Salamanna, Francesca, Melania Maglio, Gianluca Giavaresi, Stefania Pagani, Roberto Giardino, and Milena Fini. 2015. "*In Vitro* Method for the Screening and Monitoring of Estrogen-Deficiency Osteoporosis by Targeting Peripheral Circulating Monocytes." *Age (Dordrecht, Netherlands)* 37 (4): 9819. https://doi.org/10.1007/s11357-015-9819-4.

Saltel, Frédéric, Olivier Destaing, Frédéric Bard, Diane Eichert, and Pierre Jurdic. 2004. "Apatite-Mediated Actin Dynamics in Resorbing Osteoclasts." *Molecular Biology of the Cell* 15 (12): 5231–41. https://doi.org/10.1091/mbc.e04-06-0522.

Sandoval, Nathália G., Nayra S. C. Lima, Willian G. Bautz, Leticia N. Gama-de-Souza, and Karla L. A. Coburn. 2019. "Matrix Metalloproteinase 2: A Possible Role InTooth Development and Eruption." *Odovtos International Journal of Dental Sciences* 21 (1): 41–51. https://doi.org/10.15517/ijds.v0i0.35327.

Sasahara, R. M., C. Takahashi, M. C. Sogayar, and M. Noda. 1999. "Oncogene-Mediated Downregulation of RECK, a Novel

Transformation Suppressor Gene." *Brazilian Journal of Medical and Biological Research=Revista Brasileira de Pesquisas Medicas e Biologicas* 32 (7): 891–95. https://doi.org/10.1590/s0100-879x 1999000700014.

Sasaki, Takahisa, Kazuhiro Debari, and Michiko Hasemi. 1993. "Measurement of Howship's Resorption Lacunae by a Scanning Probe Microscope System." *Journal of Electron Microscopy* 42 (5): 356–59. https://doi.org/10.1093/oxfordjournals.jmicro.a051054.

Sasano, Yasuyuki, Jing-Xu Zhu, Makoto Tsubota, Ichiro Takahashi, Kazuyuki Onodera, Itaru Mizoguchi, and Manabu Kagayama. 2002. "Gene Expression of MMP8 and MMP13 During Embryonic Development of Bone and Cartilage in the Rat Mandible and Hind Limb." *Journal of Histochemistry & Cytochemistry* 50 (3): 325–32. https://doi.org/10.1177/002215540205000304.

Scherer, Sabrina, Thais Barboza de Souza, Juliana de Paoli, Claiton Viegas Brenol, Ricardo Machado Xavier, João Carlos Tavares Brenol, José Artur Chies, and Daniel Simon. 2010. "Matrix Metalloproteinase Gene Polymorphisms in Patients with Rheumatoid Arthritis." *Rheumatology International* 30 (3): 369–73. https://doi.org/10.1007/s00296-009-0974-8.

Sharma, Sudarshana M., Rong Hu, Agnieszka Bronisz, Nicolas Meadows, Tricia Lusby, Barbara Fletcher, David A. Hume, A. Ian Cassady, and Michael C. Ostrowski. 2012. "Genetics and Genomics of Osteoclast Differentiation: Integrating Cell Signaling Pathways and Gene Networks." *Critical Reviews™ in Eukaryotic Gene Expression* 16 (3): 253–78. https://doi.org/10.1615/critreveukargeneexpr.v16.i3.40.

Shi, Joanne, Mi-Young Son, Susan Yamada, Ludmila Szabova, Stacie Kahan, Kaliopi Chrysovergis, Lauren Wolf, Andrew Surmak, and Kenn Holmbeck. 2008. "Membrane-Type MMPs Enable Extracellular Matrix Permissiveness and Mesenchymal Cell Proliferation during Embryogenesis." *Developmental Biology* 313 (1): 196–209. https://doi.org/10.1016/J.YDBIO.2007.10.017.

Shibutani, Toshiaki, Kyoko Yamashita, Takanori Aoki, Yukio Iwayama, Taira Nishikawa, and Taro Hayakawa. 1999. "Tissue Inhibitors of

Metalloproteinases (TIMP-1 and TIMP-2) Stimulate Osteoclastic Bone Resorption." *Journal of Bone and Mineral Metabolism* 17 (4): 245–51. https://doi.org/10.1007/s007740050091.

Shubayev, Veronica I, Rickard Brånemark, Joanne Steinauer, and Robert R Myers. 2004. "Titanium Implants Induce Expression of Matrix Metalloproteinases in Bone during Osseointegration." *Journal of Rehabilitation Research and Development* 41 (6A): 757–66. https://doi.org/10.1682/JRRD.2003.07.0107

Shum, Lillian, Wang X., Kane A. A., and Nuckolls G. H. 2003. "BMP4 Promotes Chondrocyte Proliferation and Hypertrophy in the Endochondral Cranial Base." *International Journal of Developmental Biology* 47 (6): 423–31.

Simsa, S., O. Genina, and E. Monsonego Ornan. 2007. "Matrix Metalloproteinase Expression and Localization in Turkey (Meleagris Gallopavo) during the Endochondral Ossification Process." *Journal of Animal Science* 85 (6): 1393–1401. https://doi.org/10.2527/jas.2006-711.

Skoog, Tiina, Katja Ahokas, Christina Orsmark, Leila Jeskanen, Keiichi Isaka, and Ulpu Saarialho-Kere. 2006. "MMP-21 Is Expressed by Macrophages and Fibroblasts *in Vivo* and in Culture." *Experimental Dermatology* 15 (10): 775–83. https://doi.org/10.1111/j.1600-0625.2006.00460.x.

Sobue, T., Y. Hakeda, Y. Kobayashi, H. Hayakawa, K. Yamashita, T. Aoki, M. Kumegawa, T. Noguchi, and T. Hayakawa. 2001. "Tissue Inhibitor of Metalloproteinases 1 and 2 Directly Stimulate the Bone-Resorbing Activity of Isolated Mature Osteoclasts." *Journal of Bone and Mineral Research* 16 (12): 2205–14. https://doi.org/10.1359/jbmr.2001.16.12.2205.

Sofat, Nidhi. 2009. "Analysing the Role of Endogenous Matrix Molecules in the Development of Osteoarthritis." *International Journal of Experimental Pathology* 90 (5): 463–79. https://doi.org/10.1111/j.1365-2613.2009.00676.x.

Sorsa, Timo, Paivi Mantyla, Hanne Ronka, Pekka Kallio, Gun-Britt Kallis, Christina Lundqvist, Denis F. Kinane, Tuula Salo, Lorne M. Golub, Olli

Teronen, and Sari Tikanoja 1999. "Scientific Basis of a Matrix Metalloproteinase-8 Specific Chair-Side Test for Monitoring Periodontal and Peri-Implant Health and Disease." *Annals of the New York Academy of Sciences* 878 (1): 130–40. https://doi.org/10.1111/j.1749-6632.1999.tb07679.x.

Stamenkovic, Ivan. 2003. "Extracellular Matrix Remodelling: The Role of Matrix Metalloproteinases." *The Journal of Pathology* 200 (4): 448–64. https://doi.org/10.1002/path.1400.

Stetler-Stevenson, William G. 2008. "Tissue Inhibitors of Metalloproteinases in Cell Signaling: Metalloproteinase-Independent Biological Activities." *Science Signaling* 1 (27): re6. https://doi.org/10.1126/scisignal.127re6.

Stetler-Stevenson, William G., and Anita E. Yu. 2001. "Proteases in Invasion: Matrix Metalloproteinases." *Seminars in Cancer Biology* 11 (2): 143–52. https://doi.org/10.1006/SCBI.2000.0365.

Suda, T. 1999. "The Molecular Basis of Osteoclast Differentiation and Activation." In *Bone*, 25 (1): 160. 655 Avenue Of The Americas, New York, Ny 10010 Usa: Elsevier Science Inc.

Sulkala, M., M. Larmas, T. Sorsa, T. Salo, and L. Tjäderhane. 2002. "The Localization of Matrix Metalloproteinase-20 (MMP-20, Enamelysin) in Mature Human Teeth." *Journal of Dental Research* 81 (9): 603–7. https://doi.org/10.1177/154405910208100905.

Sulkala, Merja, Taina Tervahartiala, Timo Sorsa, Markku Larmas, Tuula Salo, and Leo Tjäderhane. 2007. "Matrix Metalloproteinase-8 (MMP-8) Is the Major Collagenase in Human Dentin." *Archives of Oral Biology* 52 (2): 121–27. https://doi.org/10.1016/J.ARCHORALBIO.2006.08.009.

Şurlin, Petra, Bogdan Oprea, Sorina Mihaela Solomon, Simona Georgiana Popa, Maria Moţa, Garofiţa Olivia Mateescu, Anne Marie Rauten, Dora-Maria Popescu, Lucian-Paul Dragomir, Ileana Puiu, Maria Bogdan, and Mihai Raul Popescu 2014. "Matrix Metalloproteinase -7, -8, -9 and -13 in Gingival Tissue of Patients with Type 1 Diabetes and Periodontitis." *Romanian Journal of Morphology and Embryology = Revue Roumaine de Morphologie et Embryologie* 55 (3 Suppl): 1137–41. http://www.ncbi.nlm.nih.gov/pubmed/25607396.

Swanson, Charlotte, Mattias Lorentzon, H. Herschel Conaway, and Ulf H. Lerner. 2006. “Glucocorticoid Regulation of Osteoclast Differentiation and Expression of Receptor Activator of Nuclear Factor-KB (NF-KB) Ligand, Osteoprotegerin, and Receptor Activator of NF-KB in Mouse Calvarial Bones.” *Endocrinology* 147 (7): 3613–22. https://doi.org/10.1210/en.2005-0717.

Takahashi, Chiaki, Zeqi Sheng, Thomas P Horan, Hitoshi Kitayama, Masatoshi Maki, Kiyotaka Hitomi, Yasuyuki Kitaura, Setsuo Takai, Regina M. Sasahara, Aki Horimoto, Yoji Ikawa, Barry J. Ratzkin, Tsutomu Arakawa, and Makoto Noda. 1998. “Regulation of matrix metalloproteinase-9 and inhibition of tumor invasion by the membrane-anchored glycoprotein RECK.” *Medical Sciences* 95 (22): 13221–26. https://doi.org/10.1073/pnas.95.22.13221

Takahashi, Kenji, Randal S. Goomer, Fred Harwood, Toshikazu Kubo, Yasusuke Hirasawa, and David Amiel. 1999. “The Effects of Hyaluronan on Matrix Metalloproteinase-3 (MMP-3), Interleukin-1β (IL-1β), and Tissue Inhibitor of Metalloproteinase-1 (TIMP-1) Gene Expression during the Development of Osteoarthritis.” *Osteoarthritis and Cartilage* 7 (2): 182–90. https://doi.org/10.1053/JOCA.1998.0207.

Takahashi, Naoyuki, Nobuyuki Udagawa, Yasuhiro Kobayashi, Masamichi Takami, T John Martin, and Tatsuo Suda. 2008. “Osteoclast Generation.” In *Principles of Bone Biology*, 175–92. Academic Press, Elsevier. http://dx.doi.org/10.1016/B978-0-12-373884-4.00029-X.

Takaishi, Hironari, Tokuhiro Kimura, Seema Dalal, Yasunori Okada, and Jeanine D’Armiento. 2008. “Joint Diseases and Matrix Metalloproteinases: A Role for MMP-13.” *Current Pharmaceutical Biotechnology* 9 (1): 47–54. https://doi.org/10.2174/138920108783497659.

Takei, Isao, Michiaki Takagi, Seppo Santavirta, Hideo Ida, Masaji Ishii, Toshihiko Ogino, Mari Ainola, and Yrj T. Konttinen. 2000. “Messenger Ribonucleic Acid Expression of 16 Matrix Metalloproteinases in Bone-Implant Interface Tissues of Loose Artificial Hip Joints.” *Journal of*

Biomedical Materials Research 52 (4): 613–20. https://doi.org/10.1002/1097-4636(20001215)52:4<613::AID-JBM5>3.0.CO;2-8.

Tang, Simon Y., Ralf-Peter Herber, Sunita P. Ho, and Tamara Alliston. 2012. "Matrix Metalloproteinase-13 Is Required for Osteocytic Perilacunar Remodeling and Maintains Bone Fracture Resistance." *Journal of Bone and Mineral Research* 27 (9): 1936–50. https://doi.org/10.1002/jbmr.1646.

Tatti, Olga, Mariliina Arjama, Annamari Ranki, Stephen J Weiss, Jorma Keski-Oja, and Kaisa Lehti. 2011. "Membrane-Type-3 Matrix Metalloproteinase (MT3-MMP) Functions as a Matrix Composition-Dependent Effector of Melanoma Cell Invasion." *PloS One* 6 (12): e28325. https://doi.org/10.1371/journal.pone.0028325.

Taz, Yi. R. Grant Rowe, Elliot L. Botvinick, Abhishek Kurup, Andrew J. Putnam, Motoharu Seiki, Valerie M. Weaver, Evan T. Keller, Steven Goldstein, Jinlu Dai, Dana Begun, Thomas Saunders, and Stephen J. Weiss. 2013. "Article MT1-MMP-Dependent Control of Skeletal Stem Cell Commitment via a β1-Integrin/YAP/TAZ Signaling Axis." *Developmental Cell* 25 (4): 402–16. https://doi.org/10.1016/ j.devcel.2013.04.011.

Tchetverikov, I., N. Verzijl, T. W. J. Huizinga, J. M. Tekoppele, R. Hanemaaijer, and J. Degroot. 2003. "Active MMPs Captured by Alpha 2 Macroglobulin as a Marker of Disease Activity in Rheumatoid Arthritis," *Clinical and Experimental Rheumatology* 21:711–18.

Teles, Ricardo, Dimitra Sakellari, Flavia Teles, Antonis Konstantinidis, Ralph Kent, Sigmund Socransky, and Anne Haffajee. 2010. "Relationships Among Gingival Crevicular Fluid Biomarkers, Clinical Parameters of Periodontal Disease, and the Subgingival Microbiota." *Journal of Periodontology* 81 (1): 89–98. https://doi.org/10.1902/jop.2009.090397.

Tetlow, Lynne C., Daman J. Adlam, and David E. Woolley. 2001. "Matrix Metalloproteinase and Proinflammatory Cytokine Production by Chondrocytes of Human Osteoarthritic Cartilage: Associations with Degenerative Changes." *Arthritis & Rheumatism* 44 (3): 585–94.

https://doi.org/10.1002/1529-0131(200103)44:3<585::AID-ANR107>3.0.CO;2-C.

Thiolloy, Sophie, Jennifer Halpern, Ginger E. Holt, Herbert S. Schwartz, Gregory R. Mundy, Lynn M. Matrisian and Conor C. Lynch. 2009. "Osteoclast-Derived Matrix Metalloproteinase-7, but Not Matrix Metalloproteinase-9, Contributes to Tumor-Induced Osteolysis." *Cancer Research* 69 (16): 6747–55. https://doi.org/10.1158/0008-5472.CAN-08-3949.

Tuter, Gulay, Bulent Kurtis, Muhittin Serdar, Aysegul Yucel, Eylem Ayhan, Burcu Karaduman, and Gonen Ozcan. 2005. "Effects of Phase I Periodontal Treatment on Gingival Crevicular Fluid Levels of Matrix Metalloproteinase-3 and Tissue Inhibitor of Metalloproteinase-1." *Journal of Clinical Periodontology* 32 (9): 1011–15. https://doi.org/10.1111/j.1600-051X.2005.00816.x.

Ubaidus, Sobhan, Minqi Li, Sara Sultana, Paulo Henrique Luiz De Freitas, Kimimitsu Oda, Takeyasu Maeda, Ritsuo Takagi, and Norio Amizuka. 2009. "FGF23 Is Mainly Synthesized by Osteocytes in the Regularly Distributed Osteocytic Lacunar Canalicular System Established after Physiological Bone Remodeling." *Journal of Electron Microscopy* 58 (6): 381–92. https://doi.org/10.1093/jmicro/dfp032.

Välimäki, Ville-Valtteri, Niko Moritz, Jessica J. Yrjans, Eero Vuorio, and Hannu T. Aro. 2006. "Effect of Zoledronic Acid on Incorporation of a Bioceramic Bone Graft Substitute." *Bone* 38 (3): 432–43. https://doi.org/10.1016/J.BONE.2005.09.016.

Varghese, Samuel. 2006. "Matrix Metalloproteinases and Their Inhibitors in Bone: An Overview of Regulation and Functions." *Frontiers in Bioscience* 11: 2949–66. https://doi.org/10.2741/2024.

Velada, Isabel., F. Capela-Silva, Flávio Reis, Euclides Manuel Vieira Pires, Conceicao Egas, Paulo Rodrigues-Santos, and Marlene T. Barros. 2011. "Expression of Genes Encoding Extracellular Matrix Macromolecules and Metalloproteinases in Avian Tibial Dyschondroplasia." *Journal of Comparative Pathology* 145 (2–3): 174–86. https://doi.org/10.1016/j.jcpa.2010.12.008.

Velleman, S. G. Sandra G. 2000. "The Role of the Extracellular Matrix in Skeletal Development." *Poultry Science* 79 (7): 985–89.

Vincenti, Matthew P., Lori A. White, Daniel J. Schroen, Ulrike Benbow, and Constance E. Brinckerhoff. 1996. "Regulating Expression of the Gene for Matrix Metalloproteinase-1 (Collagenase): Mechanisms That Control Enzyme Activity, Transcription, and MRNA Stability." *Critical Reviews™ in Eukaryotic Gene Expression* 6 (4): 391–411. https://doi.org/10.1615/CritRevEukarGeneExpr.v6.i4.40.

Vincenti, Matthew P., and Constance E. Brinckerhoff. 2002. "Transcriptional Regulation of Collagenase (MMP-1, MMP-13) Genes in Arthritis: Integration of Complex Signaling Pathways for the Recruitment of Gene-Specific Transcription Factors." *Arthritis Research* 4 (3): 157. https://doi.org/10.1186/ar401.

Visse, Robert, and Hideaki Nagase. 2003. "Matrix Metalloproteinases and Tissue Inhibitors of Metalloproteinases: Structure, Function, and Biochemistry." *Circulation Research* 92 (8): 827–39. https://doi.org/10.1161/01.RES.0000070112.80711.3D.

Vivinus-Nebot, Mylène, Patricia Rousselle, Jean-Philippe Breittmayer, Claire Cenciarini, Sonia Berrih-Aknin, Suzanne Spong, Pasi Nokelainen, Françoise Cottrez, M. Peter Marinkovich and Alain Bernard 2014. "Mature Human Thymocytes Migrate on Laminin-5 with Activation of Metalloproteinase-14 and Cleavage of CD44." *The Journal of Immunology* 172 (3): 1397–1406. https://doi.org/10.4049/jimmunol.172.3.1397.

Vu, Thiennu H., J. Michael Shipley, Gabriele Bergers, Joel E. Berger, Jill A. Helms, Douglas Hanahan, Steven D. Shapiro, Robert M. Senior, and Zena Werb. 1998. "MMP-9/Gelatinase B Is a Key Regulator of Growth Plate Angiogenesis and Apoptosis of Hypetrophic Chondrocytes." *Cell* 93 (3): 411-422. https://doi.org/10.1016/S0092-8674(00)81169-1.

Vu, Thiennu H., and Zena Werb. 2000. "Matrix Metalloproteinases: Effectors of Development and Normal Physiology." *Genes & Development* 14 (17): 2123–33. https://doi.org/10.1101/GAD.815400.

Wagner, Erwin F, and Robert Eferl. 2005. "Fos/AP-1 Proteins in Bone and the Immune System" *Immunological Reviews* 208 (1): 126–40. https://doi.org/10.1111/j.0105-2896.2005.00332.x

Wang, S.-K., Y. Hu, J. P. Simmer, F. Seymen, N. M. R. P. Estrella, S. Pal, B. M. Reid, M. Yildirim, M. Bayram, J.D. Bartlett, and J.C.-C. Hu. 2013. "Novel *KLK4* and *MMP20* Mutations Discovered by Whole-Exome Sequencing." *Journal of Dental Research* 92 (3): 266–71. https://doi.org/10.1177/0022034513475626.

Wang, Xiao dong, Yan Yiu Yu, Shirley Lieu, Frank Yang, Jeffrey Lang, Chuanyong Lu, Zena Werb, Diane Hua, Theodore Miclaua, Ralph Marcucioa, and Céline Colno. 2013. "MMP9 Regulates the Cellular Response to Inflammation after Skeletal Injury." *Bone* 52 (1): 111-19. https://doi.org/10.1016/J.BONE.2012.09.018.

Wang, Xiaofei, Jingyan Liang, Tomonari Koike, Huijun Sun, Tomonaga Ichikawa, Shuji Kitajima, Masatoshi Morimoto, Hisataka Shikama, Teruo Watanabe, Yasuyuki Sasaguri, and Jianglin Fan. 2004. "Overexpression of Human Matrix Metalloproteinase-12 Enhances the Development of Inflammatory Arthritis in Transgenic Rabbits." *American Journal of Pathology* 165 (4): 1375–83. https://doi.org/10.1016/S0002-9440(10)63395-0.

Waresijiang, Niyazi, Jungang Sun, Rewuti Abuduaini, Tayier Jiang, Wenzheng Zhou, and Hong Yuan. 2016. "The Downregulation of MiR-125a-5p Functions as a Tumor Suppressor by Directly Targeting MMP-11 in Osteosarcoma." *Molecular Medicine Reports* 13 (6): 4859–64. https://doi.org/10.3892/mmr.2016.5141.

Wei, J. C. C., H. S. Lee, W. C. Chen, L. J. Shiu, S. F. Yang, and R. H. Wong. 2009. "Genetic Polymorphisms of the Matrix Metalloproteinase-3 (MMP-3) and Tissue Inhibitors of Matrix Metalloproteinases-1 (TIMP-1) Modulate the Development of Ankylosing Spondylitis." *Annals of the Rheumatic Diseases* 68 (11): 1781–86. https://doi.org/10.1136/ard.2008.099481.

Weinbaum, Sheldon, Stephen C. Cowin, and Yu Zeng. 1994. "A model for the excitation of osteocytes by mechanical loading-induced bone fluid shear stresses" *Journal of Biomechanics* 27 (3): 339–60. https://doi.org/10.1016/0021-9290(94)90010-8

Weinger, Jesse M., and Marijke E. Holtrop. 1974. "An Ultrastructural Study of Bone Cells: The Occurrence of Microtubules, Microfilaments and Tight Junctions." *Calcified Tissue Research* 14 (1): 15–29. https://doi.org/10.1007/BF02060280.

Wilson, Carole L., Kathleen J. Heppner, Patricia A. Labosky, Brigid L. M. Hogan, and Lynn M. Matrisian. 1997. "Intestinal Tumorigenesis Is Suppressed in Mice Lacking the Metalloproteinase Matrilysin." *Proceedings of the National Academy of Sciences* 94 (4): 1402–7. https://doi.org/
10.1073/pnas.94.4.1402.

Wilson, Carole L., and Lynn M. Matrisian. 1996. "Matrilysin: An Epithelial Matrix Metalloproteinase with Potentially Novel Functions." *The International Journal of Biochemistry & Cell Biology* 28 (2): 123–36. https://doi.org/10.1016/1357-2725(95)00121-2.

Winkler, David G., May Kung Sutherland, James C. Geoghegan, Changpu Yu, Trenton Hayes, John E. Skonier, Diana Shpektor, Mechtild Jonas, Brian R. Kovacevich, Karen Staehling-Hampton, Mark Appleby, Mary E. Brunkow, and John A. Latham. 2003. "Osteocyte Control of Bone Formation via Sclerostin, a Novel BMP Antagonist." *EMBO Journal* 22 (23): 6267–76. https://doi.org/10.1093/emboj/cdg599.

Wittrant, Yohann, Sandrine Theoleyre, Séverine Couillaud, Colin R. Dunstan, Dominique Heymann, and Françoise Rédini. 2003. "Regulation of Osteoclast Protease Expression by RANKL." *Biochemical and Biophysical Research Communications* 310 (3): 774–78. https://doi.org/10.1016/j.bbrc.2003.09.084.

Woessner, J. Frederick, and Carolyn J. Taplin. 1988. "Purification and Properties of a Small Latent Matrix Metalloproteinase of the Rat Uterus." *The Journal of Biological Chemistry* 263 (32): 16918–25. http://www.ncbi.nlm.nih.gov/pubmed/3182822.

Wong, Brian R., Daniel Besser, Nacksung Kim, Joseph R. Arron, Masha Vologodskaia, Hidesaburo Hanafusa, and Yongwon Choi. 1999. "TRANCE, a TNF Family Member, Activates Akt/PKB through a

Signaling Complex Involving TRAF6 and c-Src." *Molecular Cell* 4 (6): 1041–49. https://doi.org/10.1016/S1097-2765(00)80232-4.

Woods, John F., and George Jr. Nichols. 1963. "Coliagenolytic Activity in Mammalian Bone." *Science* 142 (3590): 386–87. https://doi.org/10.1126/science.142.3590.386.

Yamamoto, Yohei, Nobuyuki Udagawa, Sachiko Matsuura, Yuko Nakamichi, Hiroshi Horiuchi, Akihiro Hosoya, Midori Nakamura, Hidehiro Ozawa, Kunio Takaoka, and Josef M Penninger. 2006. "Osteoblasts Provide a Suitable Microenvironment for the Action of Receptor Activator of Nuclear Factor-KB Ligand." *Endocrinology* 147 (7): 3366–74. https://doi.org/10.1210/en.2006-0216

Yang, P. T., X. H. Meng, Y. Yang, and W. G. Xiao. 2013. "Inhibition of Osteoclast Differentiation and Matrix Metalloproteinase Production by CD4+CD25+ T Cells in Mice." *Osteoporosis International* 24 (3): 1113–14. https://doi.org/10.1007/s00198-012-2014-x.

Yavropoulou, M. P., and J. G Yovos. 2008. "Osteoclastogenesis - Current Knowledge and Future Perspectives." *Journal of Musculoskeletal Neuronal Interactions* 8 (3): 204–16.

Ye, Shu, Gerald F. Watts, Sundhiya Mandalia, Steve E. Humphries, and Adriano M. Henney. 1995. "Preliminary Report: Genetic Variation in the Human Stromelysin Promoter Is Associated with Progression of Coronary Atherosclerosis." *British Heart Journal* 73 (3): 209–15. https://doi.org/10.1136/hrt.73.3.209.

Yokoyama, Yuko, Frank Grünebach, Susanne M. Schmidt, Annkristin Heine, Maik Häntschel, Stefan Stevanovic, Hans-Georg Rammensee and Peter Brossart. 2008. "Matrilysin (MMP-7) Is a Novel Broadly Expressed Tumor Antigen Recognized by Antigen-Specific T. Cells." *Clinical Cancer Research* 14 (17): 5503–11. https://doi.org/10.1158/1078-0432.CCR-07-4041.

Yoshihara, Yasuo, Ken'ichi Obata, Noboru Fujimot, Kyoko Yamashita, Taro Hayakawa, and Masayuki Shimmei. 1995. "Increased Levels of Stromelysin-1 and Tissue Inhibitor of Metalloproteinases–1 in Sera from Patients with Rheumatoid Arthritis." *Arthritis & Rheumatism* 38 (7): 969–75. https://doi.org/10.1002/art.1780380713.

Yoshiko, Yuji, G. Antonio Candeliere, Norihiko Maeda, and Jane E. Aubin. 2007. "Osteoblast Autonomous Pi Regulation via Pit1 Plays a Role in Bone Mineralization." *Molecular and Cellular Biology* 27 (12): 4465–74. https://doi.org/10.1128/MCB.00104-07.

Zarka-Prost-Dumont, Mylene, Frederic Jehan, Agnes Ostertag, De Vernejoul Marie-Christine, and Valerie Geoffroy. 2014. "Decrease in Expression of MMP3 in Osteoblast Protects against Bone Loss." In *Bone Abstracts in European Calcified Tissue Society Congress*, 288. Czech Republic, Prague: BioScientifica. https://doi.org/10.1530/boneabs.3.PP288.

Zhang, Keqin, Cielo Barragan-Adjemian, Ling Ye, Shiva Kotha, Mark Dallas, Yongbo Lu, Shujie Zhao, Marie Harris, Stephen E. Harris, Jian Q. Feng, and Lynda F. Bonewald. 2006. "E11/Gp38 Selective Expression in Osteocytes: Regulation by Mechanical Strain and Role in Dendrite Elongation." *Molecular and Cellular Biology* 26 (12): 4539–52. https://doi.org/10.1128/MCB.02120-05.

Zhang, Wei-Lin, Yu-Fei Chen, Hong-Zheng Meng, Jun-Jie Du, Guan-Nan Luan, Hai-Qiang Wang, Mao-Wei Yang, and Zhuo-Jing Luo. 2017. "Role of MiR-155 in the Regulation of MMP-16 Expression in Intervertebral Disc Degeneration." *Journal of Orthopaedic Research* 35 (6): 1323–34. https://doi.org/10.1002/jor.23313.

Zhang, Xu, Shuai Huang, Junchao Guo, Li Zhou, Lei You, Taiping Zhang, and Yupei Zhao. 2016. "Insights into the Distinct Roles of MMP-11 in Tumor Biology and Future Therapeutics (Review)." *International Journal of Oncology* 48 (5): 1783–93. https://doi.org/10.3892/ijo.2016.3400.

Zhang, Zhifang, and John E. Shively. 2013. "Acceleration of Bone Repair in NOD/SCID Mice by Human Monoosteophils, Novel LL-37-Activated Monocytes." Edited by Edward E. Schmidt. *PLoS ONE* 8 (7): e67649. https://doi.org/10.1371/journal.pone.0067649.

Zheng, Xuefeng, Yuanyuan Zhang, Shiming Guo, Wenming Zhang, Jinyun Wang, and Yanping Lin. 2018. "Dynamic Expression of Matrix Metalloproteinases 2, 9 and 13 in Ovariectomy‑induced Osteoporosis

Rats." *Experimental and Therapeutic Medicine* 16 (3): 1807-1813. https://doi.org/10.3892/etm.2018.6356.

Ziober, Barry L., Mary Ann Turner, Joel Michael Palefsky, Michael J. Banda, and Richard H. Kramer. 2000. "Type I Collagen Degradation by Invasive Oral Squamous Cell Carcinoma." *Oral Oncology* 36 (4): 365–72. https://doi.org/10.1016/S1368-8375(00)00019-1.

In: A Closer Look at Metalloproteinases ISBN: 978-1-53616-517-3
Editor: Lena Goodwin

Chapter 7

TECHNIQUES TO IDENTIFY AND MEASURE TYROSIDE PHOSPHATASE INHIBITORS

Marisa Cabeza*[*]*, PhD
Department of Biological Systems. Metropolitan University-Campus Xochimilco Mexico City, Mexico

ABSTRACT

Protein tyrosine phosphatases (PTPs) are important targets that are known to play a key role in the development of chronic degenerative diseases such as obesity, diabetes, and some neurological diseases.

To date, different strategies have been developed to produce both reversible and irreversible inhibitors of PTP activity; above all, selectivity has been sought. This has proven to be a challenge for researchers in the field of medicinal chemistry due to the high conservation and cationic nature of the active sites of several PTP. However, small electrophilic molecules with inhibitory activity have been developed to target active sites with high activity and selectivity. Currently, the challenge lies in the design of allosteric and covalent inhibitors to modulate the activity of PTP. The strategy of allosteric and covalent inhibition has generated some successful high-activity small molecules for PTP1B.

[*] Corresponding Author's E-mail: marisa@correo.xoc.uam.mx.

In this regard, recent advances have led to some specific class of inhibitors. However, none of them have exhibited true selectivity towards a given PTP.

Keywords: protein tyrosine phosphatases, allosteric and covalent inhibitors, obesity, diabetes, neurodegenerative diseases, relaxation time constant, starting constant of enzymatic inhibition, inactivator molecules

1. Introduction

Human health has been a subject of great concern for governments over the past two centuries. Thus, many researches have focused on discovering new drug targets to design specific therapeutics with a view to reduce the damaging side effects of current treatments. In order to accomplish this objective, studies have been carried to study the drugs-mechanisms of action in different targets. Receptor modulators and enzyme inhibitors have been items in these investigations, which has underpinned the importance of enzymology for pharmaceutical chemists as well as pharmacologists to evaluate the different aspects of drug discovery.

Historically, the initial studies on enzymes can be traced back to the 18th century with the work of Lazzaro Spallanzani. This physiologist observed that the presence of ferments into the digestive system was capable of converting food into liquids in the stomach and intestine. Subsequently, several works in digestion physiology demonstrated the conversion of certain types of food into simple sugars and amino acids through hydrolysis, thereby stimulating the curiosity to purify and ascertain the nature of these ferments.

First researches about enzyme purification are attributed to Payen and Persoz, who, precipitated amylase from a liquid extract of malt treated with ethanol in the year 1833. Meanwhile in 1836 Schwann extracted pepsin from the stomach-wall acid fraction and made the first demonstration of its role in catalysis. Later, Pasteur studies proved that fermentation is an essential part in the living beings. Thereafter, in 1860, Berthelot mixed ethanol with yeasts culture, evidencing the presence of ferment in the formed precipitate that was able to convert glucose in fructose. Subsequently, Hans and Edouard Buchner extracted a juice (free of cells) from yeasts, which totally fermented the sugar present. All those ferments went on to be known as enzymes.

Moreover, by this time, Liebig's theory elucidated that enzymes were chemical substances with constant configuration changes, whereas Emil Fisher postulated these should be endowed with a high degree of specificity. In 1900, Bertrand analyzed the remaining enzyme dialyzable fractions and discovered that enzymes bind to different factors, which he referred to as coenzymes. Correspondingly, Sörensen in 1909 determined the effect of pH on catalytic reactions mediated by enzymes, thus reinforcing the importance of pH in these catalytic processes.

However, Henri and Brown demonstrated in 1902 that enzymes require a binding substrate in order to perform their catalytic action. They based this concept on a curve obtained when the velocity of reaction against substrate concentration were plotted. Eleven years later, Michaelis and Menten rediscovered the equation derivate by Henri's experiment. In conjunction with Briggs and Haldane, they introduced the concept of enzymatic kinetics that continues to be used till date to explain enzymes kinetic properties. Meanwhile in 1960 Daziel, Alberty, Hearon and others posited that enzymes do not always adhere to the Michaelis-Menten kinetics. To that end, Cleland demonstrated a procedure to describe the substrates of multi-reactive stable-state enzymes. Finally, in 1965 Monod, Wyman, and Changeux proposed a kinetic model for regulatory enzymes (allosteric enzymes).

During the 1930's, Northrop and coworkers crystallized pepsin, trypsin, and chymotrypsin, demonstrating the chemical nature of enzymes as proteins. Hundreds of new enzymes were purified, crystallized, and identified during the 1940s' and 1950s. New chemical and physical techniques were used for this purpose, and biochemists focused on determining their activity and regulation. In 1960, ribonuclease was sequenced, whereas the enzyme was synthesized in 1969. Notably, one of the most important discoveries of the 20th century was the identification of the reverse transcriptase, an enzyme capable of enabling DNA sequencing from RNA within the laboratory. This denotes an indirect method of determining the sequence of a protein and hinges on the genetic code. This discovery led to the creation of many molecular biology techniques that are widely applied up to date for the development of diagnostic procedures, vaccines, drugs, drug-screening system, and enzymes replacement protocols adopted in gene therapy.

2. Identification of an Enzyme Activity as a New Drug Target

For identifying a new therapeutic target, it is necessary to gather plenty of evidence to demonstrate its structure and function. For instance, the activity of many phosphatases is currently been investigated [1, 2] because they are believed to be involved in different diseases. This notion is premised on the results that were previously reported by various researchers concerning the activity of phosphatases as dephosphorylating agents. This activity could be crucial in important signaling pathways, initiated by kinases [3-5].

While the functionality and characterization of kinases has been described over the past three decades, there is paucity of research when it comes to the activity of phosphatases. Early researches on protein tyrosine phosphatase (PTPase) were developed in mutant sensitive to temperature Rous sarcoma virus. This model [6] demonstrated the increase in phosphotyrosine levels at a temperature of 36°C and exhibited tyrosine kinase activity at this temperature. However, increasing the temperature to only 41°C reduced the levels of phosphotyrosine, thereby indicating the presence of tyrosine phosphatase activity. Similar results were documented when labeled phosphorus was incorporated into the tyrosyl residues of membrane proteins that were treated with epidermal growth factor. In turn, this evidenced the presence of tyrosine kinases coupled with the epidermal growth factor receptor, whereas the slow release of the labeled phosphorus was indicative of a tyrosine phosphatase activity [7].

With this recognized activity of tyrosine phosphatases as the background, Tonks and coworkers [8] in 1988 purified the main protein tyrosine phosphatases from human placenta. Catalytic subunits of phosphatases 1 and 2 obtained from placenta were isolated by reverse-phase fast protein liquid chromatography. Subsequently, these fractions were purified by chromatography on polylysine-sepharose. [9] Using this method, two different catalytic subunits were identified in isolated tyrosine phosphatases. These subunits were specific and exhibited a high binding affinity for the phosphotyrosyl residues, resulting in their dependence on the presence of sulfhydryl groups in the reaction to get activated. Importantly, these groups bonded with high affinity to a highly reactive cysteinyl residue that was also present in these enzymes. [8] Following these methods, Tonks et al. reported the presence of two different phosphatases PTP 1A and PTP 1B in human placenta tissue to reveal different biochemical and pharmacological properties.

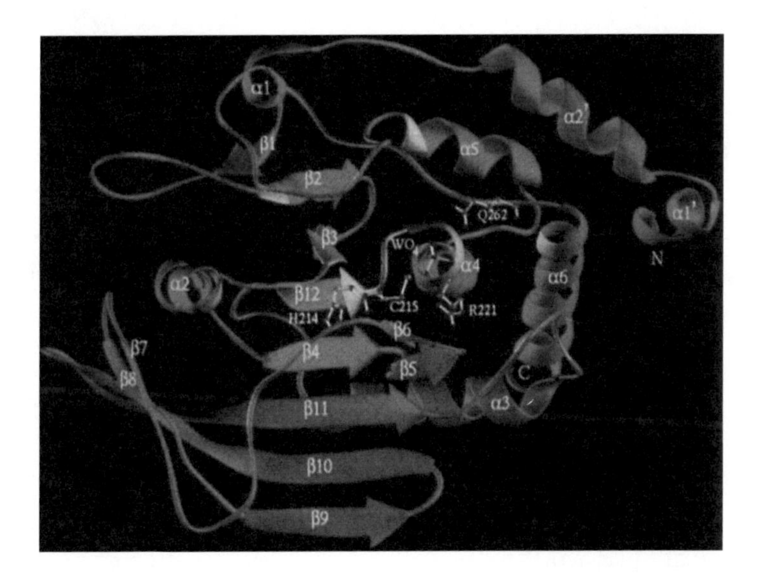

Figure 1. Secondary structure of PTP 1B, catalytic site and invariant residues (yellow). Helices α and β are illustrated in this figure as tungstate ion (WO) used as an inhibitor of activity of this phosphatase. This ion is located linearly with α4; side chains of His^{214}, Cys^{215}, and Gly^{262} are present as well. Catalytic site is located on the COOH terminal of the central region. Abbreviations: A, Ala; C, Cys; D, Asp; E, Glu; F, Phe; G, Gly; H, His; I, lie; K, Lys; L, Leu; M, Met; N, Asn; P, Pro; Q, Gin; R, Arg; S, Ser; T, Thr; V, Val; W, Trp; and Y, Tyr. W04, tungstate. Figure sourced from Barford D, Flint AJ Tonks NK [10].

After the discovery of tyrosine phosphatases 1A and 1B, their crystal structures were subsequently investigated. [10, 11] These enzymes, which catalyze the dephosphorylation of phosphotyrosyl residues, are characterized by homologous catalytic domains. Crystal structure analysis of PTP 1B indicated a 37 kilodalton protein containing 321 amino acid residues. In addition, it demonstrates a unique domain with the catalytic site located in a surface cleavage along with the phosphate recognition site at the amino terminus of the α-helices. This catalytic site consists of a sequence of 11 residues that, in turn, contain cysteine and arginine. The invariant position of the cysteine residue within the phosphate-binding site imparts a neutrophil role in this reaction (Figure 1).

3. Physiological Role and Mechanism of Action of Phosphatase 1 B

Tyrosine phosphatase, PTP 1B, is found in the endoplasmic reticulum of insulin target cells, such as hepatocytes, muscle cells, and fats. This enzyme regulates the deactivation of insulin and leptin receptors, whereas dephosphorylating tyrosine residues shows kinase activity. In 2008, Picardi and coworkers used a designed antisense oligonucleotide to block the expression of PTP 1B in the hypothalamic nuclei around the third ventricle in both normal and obese rats. This oligonucleotide was administered to the rats by infusion; according to the results, food intake, body weight, and fat accumulation after meals decreased in both types of rats. In addition, the findings indicated that both leptin and insulin actions improved in normal and obese rats [12, 13].

The physiological role of PTP 1B in fat metabolism appears to be stimulating lipogenesis, given that the dephosphorization of the leptin-tyrosine-kinases-receptor causes an increase of fatty acid synthesis and a decrease in β-oxidation. This is in contrast to the function of the leptin-receptor activated by tyrosine-kinase activity, thus decreasing the fat metabolism. [3] This idea is premised on the fact that PTP 1B regulates the tyrosine kinase signaling pathway following leptin or insulin stimulation.

The mechanism of action of PTP 1B in the leptin signaling pathway has been recognized. [14]. Leptin binds to its receptor and signaling pathway begins to allow phosphorylation of Janus-kinase 2, which then triggers to STAT3. Thereafter, activated STAT3 translocates to the nucleus in order to induce gene response, which decreases malonyl CoA and fatty acid synthesis. Additionally, STAT 3 increases β-oxidation, thus resulting in a reduction of fat accumulation. On the other hand, when PTP 1B dephosphorylates the leptin receptor, an accumulation of fat is observed in these tissues [16, 17]. This, in turn, suggests that the inhibition of PTP 1B could be considered as a palliative option to improve obesity and insulin resistance [15, 16].

In sugars metabolism, PTP 1B assumes significance also because it can induce the insensibility to insulin signaling. Studies have demonstrated that the inhibition of this enzyme in diabetic and non-diabetic mice prevents weight gain while increasing the sensitivity of insulin. [14] It was previously demonstrated that the absence of PTP 1B in knockout mice increases insulin sensitivity in the liver, thereby decreasing the gluconeogenic genes PEPCK and G-6-Pase. The production of glucose by the liver is inhibited accordingly, improving glucose tolerance. Deletion of hepatic PTP 1B also decreases cholesterol and triglyceride levels, and the expression of SREBPs, FAS and ACC genes improves mice health [17].

Thus, data in general indicates that the inhibition of this enzyme would be beneficial in medical ailments such as insulin resistant syndrome, obesity, and type 2 diabetes. This has prompted the development of several new drugs that are aimed at activating leptin and insulin mechanisms of action. Therefore, some 2,4-thiazolidinediones derivatives have been designed for this purpose. However, there are still on phase II of clinical trials.

4. Evaluation of Enzyme Inhibitors

To evaluate new molecules as enzyme inhibitors aimed to develop a new drug, it is paramount to comprehend the relationship between the drug and new target enzyme. It is also important to understand competitive, noncompetitive, allosteric and uncompetitive inhibition, slow binding, tight binding, and reaction stoichiometry. These data are derived from the *in vitro* evaluation of enzyme-inhibitor interactions.

The first experiments reveal the presence of phosphatases in cell membranes involving human epidermoid carcinoma cell membranes that are rich in epidermal growth factor receptors [8] and can incorporate labeled phosphorus into the tyrosyl residues of different membrane proteins. This was observed in presence of the epidermal growth factor (EGF), thereby indicating the presence of kinase enzymes. However, membrane-associated dephosphorylation reactions were not impacted by EGF, thus indicating the presence of phosphatases [8].

PTPs activity was demonstrated later in the membranes of different tissues and cell lines rich in these proteins [3]. Some of these tyrosine phosphatases were purified and characterized using human placenta. In this tissue, a 35 KD tyrosine phosphatase was found in both the soluble and particulate fraction. Activity reported for the soluble fraction was of 2000 units/g protein, whereas the kinetic parameters indicated a velocity rate of Pi-releasing of 45 μmol/min/mg. [19] This value was shown to be twice as higher than that was previously reported for type 1 and 2A serine/threonine phosphatases. [19] Two subtypes, A and B of these PTPs, have been elucidated in these membranes. [8] Subtypes A and B meanwhile show a different response to polyanionic and polycyclic compounds. While EDTA, spermine, spermidine, and basic myelin activate those of subtype B, this reaction is not observed in the case of subtype A. Moreover, subtype B could be inhibited with low concentrations of heparin and glutamate/tyrosine copolymers, which is not the case with type A.

The amino acid sequence of subtype B has been also informed. [20] It consists of a 321 amino acid single-chain protein with N-terminal N-acetylated methionine and an unusually proline-rich C-terminal region

(Figure 1). The presence of cysteinyl residues 121 and 215 in subtype B, are conserved among all PTPs members, which indicates that these are paramount for its catalytic function. [6] Structural and mutational studies have identified the presence of a common catalytic in all members of the family of PTPs [21]. This catalytic site is formed by a β-sheet of 250 amino acids located parallel to the α-helices having a β-loop-α catalytic loop, which contains motif characterizing the PTP (I/V) family HCXXGXXR (S/T), as shown in Figure 1. This loop comprises of 11 amino acid residues, of which cysteine and arginine residues are paramount for its catalytic activity. In addition, the phosphotyrosine (pTyr) loop contains a tyrosine residue (Tyr46 in PTP 1B) that defines the depth of the active site and also contributes to the specificity of substrates containing pTyr46.

4.1. Mechanism of PTP 1B Catalysis

The catalytic function of the PTP 1B is achieved through the cysteine thiolate (pH 7.4) located on its active site. The cysteine thiolate acts as a nucleophile in order to initiate the hydrolysis of the phosphate monoester of the substrate's tyrosine (pTyr), until the formation of a phospho-cysteine (Figure 2) [22, 23].

In this reaction, the pTyr residue of the substrate is bound by a hydrogen bond to the guanidine that is found in the side chain of the arginine residue of the active site positively charged, to form phospho-cysteine. (Figure 2) Aspartic acid, which is present in the WPD loop, also participates in the process, adding protons to the tyrosyl group of the substrate. Finally, the glutamine located in the G loop of the active site coordinates with aspartic acid through hydrogen bonds to create an active water molecule. (Figure 2) This reacts with phospho-cysteine to release the inorganic phosphate, once again forming cysteine thiolate Figure 2 [22].

The knowledge of the structural determinants of the catalytic site makes PTP 1B an important therapeutic target to improve the study of diseases related to this enzyme. The development of selective inhibitors can be turned into a reality if it is considered that only some part of the inhibitory molecules interacts with the presence of residues in the active site of this enzyme, while the remaining part establishes contact with external residues, as has been revealed for protein kinases [24]. Previous studies have

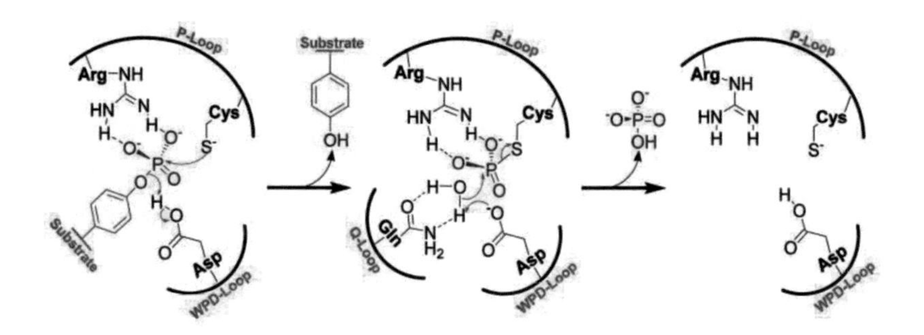

Figure 2. Catalysis mechanism achieved by tyrosine phosphatase 1B to pTyr substrates. Figure sourced from Viswanatharaju K, et al. [22].

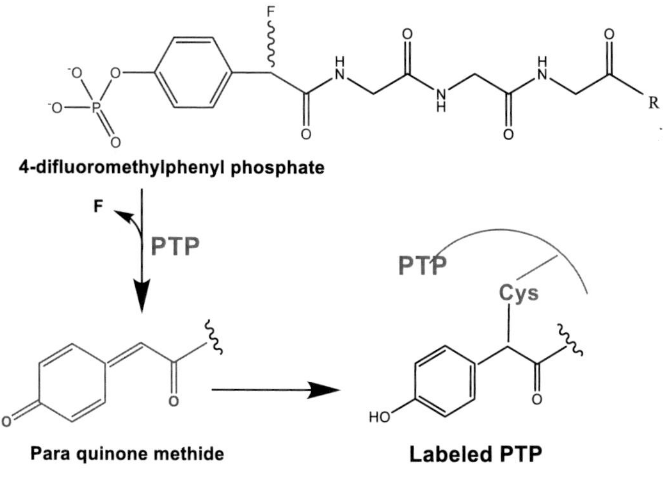

4-difluoromethylphenyl phosphate

PTP

Para quinone methide

Labeled PTP

Figure 3. 4-Difluoromethyl phenyl phosphate bound to a peptide and used as a direct PTP 1B probe to construct a covalently labeled PTP 1B.

demonstrated that the use of PTP probes can be very useful to show covalent bond formation with different compounds. For instance, adding the reactive quinone methide motif to this probe can be covalently linked to the cysteine residue of the active site of the PTP 1B. Since the quinone methide is an electrophile, it can alkalize any nearby nucleophile, such as the thiol group of the cysteine found in the active site of the PTP 1B. Formation of the covalent bond can irreversibly inactivate this enzyme by releasing the inorganic phosphate [25, 26]. Figure 3 illustrates the formation of PTB 1B marked by this probe that selectively targets PTP-1B [27].

4.2. Recent Advances in the Development of Covalent Inhibitors

Covalent inhibitors are known to exhibit different advantages, such as their high activity and prolonged interaction with their protein targets. However, this interaction could induce side effects and toxic effects as electrophilic agents affect the functioning of cytochrome P450 [28]. Despite this, many medications such as aspirin, penicillin, lansoprazole, and omeprazole, which induce covalent modification of their target proteins, became best sellers in the United States in the year 2009 [29].

Covalent inhibitors of PTP do not exhibit a kinetic equilibrium for the inactivation of the enzyme as in a reversible reaction, because this inhibition depends on the concentration of the inhibitor as well as the time of formation of the covalent bonds. These blockers neutralize the target protein, as opposed to reaching the equilibrium of the reaction. Thus, in order to study the type of inhibitor in question and decipher its affinity for a given target, it is important to first determine whether the inhibition of the PTP by each compound is slowly reversible or irreversible.

Reversibility of an enzymatic inhibition can be determined by measuring the recovery of the enzyme activity after rapid formation of the enzyme-inhibitor complex. If the inhibition process is found to be reversible, the time-product plot will be linear, as depicted in Figure 4. [30] However, if the inhibition is irreversible or very slowly reversible, the plot will assume the form of a curve. The curvature of the progress graph will reflect a slow recovery of the activity of the enzyme, given that the inhibitors dissociate very slowly from the target enzyme, as shown in Figure 4 [30].

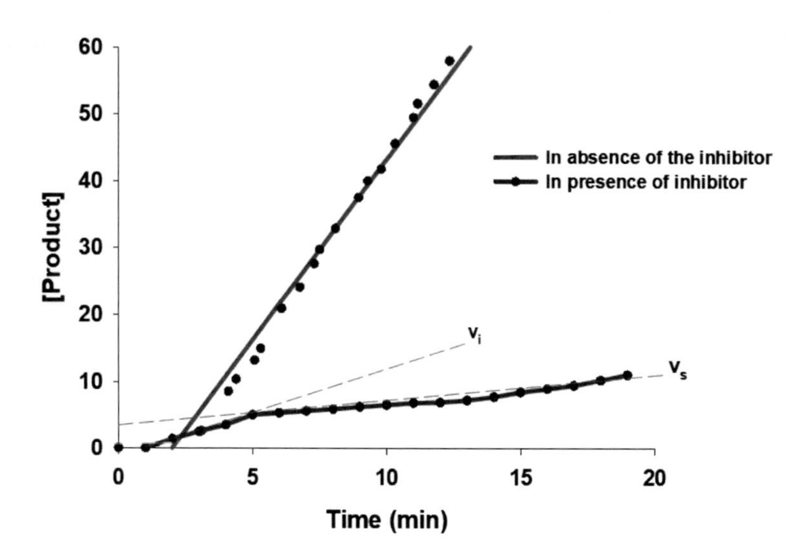

Figure 4. Plot of variation of enzymatic activity over time in the absence or presence of a slow binding inhibitor at a fixed concentration of both the inhibitor and the substrate.

This figure illustrates the evaluation of the affinity of the inhibitor; however, the classical method of stable-state is unsuitable for these slow binding inhibitors, since there is a possibility of underestimating the activity of this blocker by the target enzyme. Figure 4 shows a significant curvature (blue plot) caused by the transition from initial to the steady state velocity. Thus, the enzyme activity at a fixed concentration of the slow binding inhibitor could be described by the following equation:

$$[P]=v_s t+\frac{v_i\,[1-\exp(-K_{obs}\,t)]}{K_{obs}} \quad (1)$$

[P] Product concentration

v_s = Velocity measured near of the end of the curve

v_i = Velocity measured from the slope of the early time points

K_{obs} = Constant for conversion from vi phase to the steady state (vs) velocity phase.

T = time.

Determining the value of starting constant of enzymatic inhibition (K_{obs}) is useful because it projects the behavior of the inhibitor with its target enzyme (E). To illustrate, as shown in Figure 4, it is possible to easily verify (by means of the behavior of the plot) whether it pertains to a reversible, slow-binding, or irreversible inhibitor at a specific concentration of the inhibitor (I). Covalent inhibitors first bind reversibly to the enzyme in order to form an *EI* complex. However, by allowing the formation of covalent linkages with the active site of the enzyme, the group of interest is released, a mentioned above for PTP example (see Figure 2). This covalent bond produces a very small return speed value (K_2) for the recovery of substrate as well as enzyme in their stable state. In this case, the value of v_s approaches zero in all concentrations used of the inhibitor, whereas K_{obs} is depicted as a hyperbole in the plot, with the intersection with the y-axis in close proximity to the origin. In this case, K_{obs} will have a value equal to K_1:

A plot that describes the behavior of covalent inhibitors whose mechanism of action is predicated on the inactivation of enzymes shows that K_{obs} is a function of [*I*], regardless of how high its concentration is. In addition, this type of inactivators disallows any dissociation of the *EI* complex. Therefore, the graph of K_{obs} against [*I*] shows that the K_{obs} intercepts with the y-axis at zero (Figure 6).

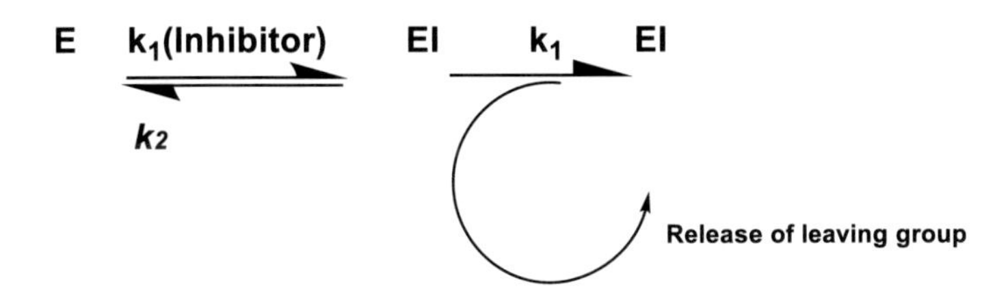

Figure 5. Reaction mechanism for covalent modification of an enzyme by a selective inhibitor with high affinity for the active site.

Therefore, on the basis of Michaelis-Menten equation, the maximum rate of the inactivation of an enzyme at an infinite concentration of inactivator is described as follows:

$$k_{obs} = \frac{k_{Inact}\,[I]}{K_I + [I]} = \frac{k_{Inact}\,[I]}{1 + (K_I + [I])} \tag{2}$$

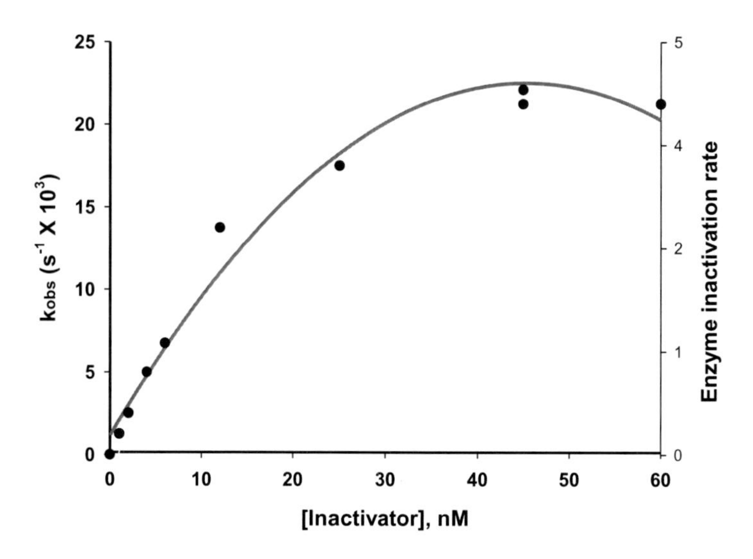

Figure 6. Plot describing the behavior of K_{obs} as a function of the enzyme inactivator's concentration *[I]* in a two-step mechanism. First step involves the reversible binding of the inhibitor to the enzyme. Meanwhile the second step involves the formation of covalent bonds that could still be lightly reversible. However, the inactivators never produce a dissociation of the covalent complex *EI*, which implies that the intersection in y-axis of the graph would be equal to zero.

Analogous to the term K_m from the Michaelis-Menten Plot, K_I is defined as the concentration of the inactivator that yields the rate of inactivation equal to $\frac{1}{2}k_{inact}$. In addition, at any given inactivator concentration, the half-life for inactivation can be defined as τ/k_{obs}. Meanwhile the time required for the decay of the [*EI*] has been referred to as the relaxation time constant (τ)

$$\tau = \frac{1}{k_2} \tag{3}$$

Thus, a velocity equation for the first-order reaction could be proposed, taking into consideration, a unit of time ½ K_{inact} as:

$$[EI]_t = \frac{1}{2}\,[EI]_0 \tag{4}$$

$[EI]_0$ Enzyme-inactivator complex formation in the time zero
$[EI]_t$ Enzyme-inactivator complex formation in a given time

So,

$$\ln\left(\frac{[EI]_t}{[EI]_0}\right) = \ln(0.5)=-0.6931 \tag{5}$$

Additionally, a half-life of inactivation can be defined at infinite concentration at ½ time equal to:

$$\text{Ln}(0.5)=-0.6931=-k_2 t_1/2 \tag{6}$$

Or,

$$\ln\left(\frac{[EI]_t}{[EI]_0}\right) = \ln(0.5)=-0.6931 \tag{7}$$

And,

$$t_{1/2} = \frac{0.6931}{k_2} \tag{8}$$

$$^{*}t_{1/2} = 0.6931\ \tau$$

*This equation is used to determine the inactivation efficiency value of an inhibitor.

4.2.1. Nonspecific Affinity of Covalent Inhibitors

The affinity of covalent inhibitors for an enzyme is an inherent property of each one of them. Such blockers modify the appropriate nucleophiles that are found both inside and outside the enzyme molecule, so as to facilitate the formation of the covalent bonds are formed. For this reason, their lack of specificity for a target enzyme becomes evident; this fact has significantly curtained its usage as reliable drugs.

However, it is necessary to consider this as a two-step reaction. The first step involves a reversible reaction that only occurs in the active site of the enzyme and probably with a very weak electrophile. This reaction can appropriately guide the weak electrophile into the target enzyme in order to make it react with the suitable nucleophile; as a result, covalent bonds are formed. This strategy can add some selectivity for the target enzyme, but does not increase its affinity for it. There are a number of drugs that follow this strategy. For example, aspirin selectively acetylates a serine residue within the active site of COX1 and COX2 by irreversibly inactivating the activity of these enzymes. Their antithrombotic and anti-inflammatory effects are derived from this reaction.

The lack of affinity of a given drug for the target enzyme can cause several toxicities. It is for this reason that therapeutics currently focuses on identifying specific drugs for specific targets in order to accentuate their efficacy by decreasing their adverse effects. Therefore, selective deactivators must be carefully tested, particularly in long-term and chronic drugs.

Another latent risk for the development of covalent drugs is the immunological reactions that may possibly occur after the formation of the drug-enzyme complex. Due to the fact that the immune system is far from responding to xenobiotics of molecular weight less than 1000 Daltons, new technologies must depend on this data to develop covalent drugs. The immunological reaction for these complexes could vary from an anaphylactic reaction to hemolytic anemia or organ damage. In addition, blood transporter proteins could covalently bind to these nonspecific drugs, thus causing unexpected adverse immune reactions for the patient. Reasons such as these have deterred the development of drugs linked covalently despite being very effective and highly estimated in the market.

4.2.2 Covalent PTP Inhibitors

Considering the reactivity of the groups present in a covalent inhibitor to induce modifications in the amino acid residues of the active site of the PTP enzymes, Ruddraraju and Zhang in 2017 [31] grouped them into eight classes:

4.2.2.1. Oxidizing Agents

These covalent inhibitors inactivate PTP activity to convert the cysteine residue of the active site into a sulfenyl or disulfide amide [32-34]. Although oxidation of the thiol group from cysteine to sulfenic acid (SOH) or sulfanilamide is a reversible reaction, the formation of sulfinic acid (SO_2H) or sulfonic acid (SO_3H) through over oxidation of this same group induces an irreversible loss of the catalytic activity of PTP.

Meanwhile Bhattachara and coworkers [35] previously demonstrated that hydroperoxides induce a covalent modification of the PTPB enzyme. Additionally, other oxidizing agents such as NO have also been found to be capable of inactivating PTPs through S-nitrosylation of protein thiols [36].

In addition to these oxidants, it has been previously reported that peroxynitrite, a biological oxidant, irreversibly blocks the activity of three known human PTPs. These reactions are induced by the oxidation of the thiol that is present in the active site to either sulfinic or sulfonic acids. These oxidation products cannot be reduced by biological thiols, such as dithiothreitol (DTT) [37].

4.2.2.2. Phosphates y Phosphonates

The addition of an electrically active isostere of the phosphonate to the Tyr residue has been known to produce selective and irreversible inhibitors of PTPs. [38-40] PTP active site-directed mutagenesis experiments [41] have previously indicated the formation of a covalent adduct between the isosterers, such as the α-bromobenzylphosphonates (α-BBP) and the Cys residue of the active site. [40] Furthermore, its selectivity for this active site has been attributed to its structural similarity to that of the phosphotyrosine. Thus, it is possible to explain its action mechanism, given that the electrophilic group (α-BBP) is juxtaposed with the nucleophile of the active site of PTP and. Consequently, the activity of the enzyme is irreversibly inhibited.

4.2.2.3. α-Halo Carbonyls

Among several existing α-halo carbonyls, it has been shown that iodoacetate is a potent inactivator of the cytoplasmic domain of PTP. This can be explained by the presence of positive charges of the residues inside the active site of this enzyme. In the PTP of *Yersinia* bacteria, the iodoacetate showed a selective and stoichiometric reactivity with the active site cysteine (Cys403) of PTP. Thus, α-halo carbonyls denote compounds displaying an electrophilic center that reacts with nucleophilic residues of proteins [10].

4.2.2.4. α,β-Unsaturated Compounds

The most common reactions involved in the covalent modification of proteins are the Michael's additions. Functional groups that mostly experience these additions are α,β-unsaturated carbonyls, vinyl sulfones, vinyl sulfonates, quinones, alkyl amides, propargylic acid derivatives, as well as cyanoacrylates [42]. Thus, Michael's addition of the thiol of cysteine to these groups can possibly block cysteine residues of the active site of PTP, which, in turn, can selectively and irreversibly inhibit the enzyme activity [43].

It has been previously reported that acrolein is a time-dependent inactivator of PTPB. This fact assumes significance because the sulfur atom from the cysteine 215 residue is known to reacts with the presence of double bond in the β-carbon of acrolein *via* a Michael's addition [44], as shown in Figure 7.

4.2.2.5. Carbonyl Compounds

Peptide aldehydes are known to react with nucleophiles of the active site of proteases to form either hemiacetal or hemithioacetal covalent bonds, which then simulate the transition state of a normal enzymatic reaction.

Pei and coworkers [46] demonstrated the formation of a covalent imine adduct (E-I) between the guanidine group of the PTPB active site Arg 221 and the aldehyde group of the inhibitor. Conjugate addition at the β-carbon by a nucleophile produces an enamine (E-I).

In contrast, and according to molecular modeling studies, Shim and coworkers [45] observed that some formylchromone derivatives were potent and selective inhibitors of the PTP 1B activity, owing to the formation of thioester bonds. Because the S atom of Cys215 was oriented in the correct position towards the carbonyl of these derivatives' aldehyde, the nucleophilic attack was made possible.

4.2.2.6. Isothiocyanates

Recently, Lewis et al. [45] reported that isothiocyanates induce covalent modification of PTP 1B in the thiol of the cysteine residue at its active site. For this purpose, one of the best-known isothiocyanates is 6-methylsulfinylhexyl isothiocyanates (6-HITC), which is the primary wasabi compound. This compound successfully improved neuritogenesis in PC12 cells by inhibiting PTP 1B. However, its action mechanism is yet to be reported. Despite this, 6-HITC does show potential as a basis for the development of drugs that can be used to treat neurodegenerative diseases.

4.2.2.7. Epoxides

Because epoxides (oxiranes) are cyclic ethers with a ring of three atoms, they contain electrophilic carbon atoms that can be attacked by a variety of nucleophiles. This characteristic makes them interesting as inhibitors of PTPs activity. An example of these compounds is the fosfomycin, 1,2-epoxy-3-(nitrophenoxy)propane (EPNP), and (R)-(-) and (S)-(+)-2-(benzyloxymethyl)oxiranes. Their activity has been shown to be irreversible covalent inactivators of PTPs. Studies have also revealed that EPNP bind covalently to Cys62 and Cys145 in low molecular weight PTPs through an allosteric type inhibition [46].

4.2.3. Other PTP Inactivation Mechanisms

So far, different mechanisms of action of covalent inactivators of PTP have been described. However, the analysis published by Ruddraraju and Zhang in 2017 also mentions other mechanisms of action for inhibitors of PTP activity [22].

For example, fluorogenic 4-(aminosulfonyl)-7-fluoro-2,1,3-benzoxadiazole was found to be able to allosterically inhibit PTP 1B by covalently modifying the Cys 121 fragment. This type of inhibition was also determined to operate with 1,2-Naphthoquinone, which also showed some interactions with the His-25 and Cys-215.

Both these examples demonstrate that the presence of cysteine residues in bags located outside an active site can also be used for the design of selective PTP inhibitors.

5. Conclusion

PTP are very important therapeutic targets because they play a pivotal role in the development of chronic degenerative diseases, such as obesity, diabetes, as well as some neurological diseases.

To date, different strategies have been formulated to produce both reversible and irreversible inhibitors of PTP activity; above all, selectivity has been actively sought. However, this has posed a challenge for researchers in the field of medicinal chemistry due to the high conservation and cationic nature of the active sites of several PTP. However, small electrophilic molecules with inhibitory activity have been developed to target active sites with high activity and selectivity. Currently, the challenge is to design allosteric and covalent inhibitors in order to modulate the activity of PTP. The strategy of allosteric and covalent inhibition has been reported to produce some successful small high-activity molecules in the case of PTP 1B.

There are advantages of preferring covalent inhibitors over traditional non-covalent inhibitors. Since low doses are required to exhibit the desired pharmacological effect and generate stable and durable bonds, they are resistant to mutations. Although interactions with other PTP outside the therapeutic target are indeed a major concern, careful and systematic analysis of the drugs developed could catalyze the design of a selective drug for each class and encourage more ambitious endeavors for each subclass of PTP. In this regard, recent advances have been made since some of the inhibitors designed have shown to be inhibitors of a specific class. However, none of them has displayed true selectivity towards a given PTP.

The approach to induce specificity for the covalent inhibition of PTP entails designing bifunctional molecules with a specificity objective and another one that is capable of creating a covalent bond with the preferred PTP, for example, the reaction of the inhibitor with an active site residue. Although the structures of the active sites are highly conserved among several PTPs, it must be noted that the surfaces of these proteins are very varied. Therefore, the strategy also involves directing the molecules of the inhibitors into the non-conserved surfaces.

References

[1] Russo, L.C., Farias, J.O., Ferruzo, P.Y.M., Monteiro, L.F. and Forti, F.L. (2018). *Revisiting the roles of VHR/DUSP3 phosphatase in human diseases*. Clinics (Sao Paulo). Copyright © 2018 CLINICS.

[2] Kim, M., Morales, L.D., Jang, I.S., Cho, Y.Y. and Kim, D.J. (2018). Protein Tyrosine Phosphatases as Potential Regulators of STAT3 Signaling. *Int J Mol Sci*. 19.

[3] Koren, S. and Fantus, I.G. (2007). Inhibition of the protein tyrosine phosphatase PTP1B: potential therapy for obesity, insulin resistance and type-2 diabetes mellitus. *Best Practice & Research Clinical Endocrinology & Metabolism*. 21, 621-640.

[4] Yang, C.F. (2018). Clinical manifestations and basic mechanisms of myocardial ischemia/reperfusion injury, Ci Ji Yi Xue Za Zhi. Copyright: © 2018 *Tzu Chi Medical Journal*., pp. 209-15.

[5] Qi, X.M., Wang, F., Mortensen, M., Wertz, R. and Chen, G. (2018). Targeting an oncogenic kinase/phosphatase signaling network for cancer therapy. *Acta Pharm Sin B*. 8, 511-517.
[6] Sefton BM, Hunter T, Beemon K, Eckhart W. Evidence that the phosphorylation of tyrosine is essential for cellular transformation by Rous sarcoma virus. *Cell*. 1980; 20(3):807-16.
[7] Carpenter, G., King, L. and Cohen, S. (1979). Rapid enhancement of protein phosphorylation in A-431 cell membrane preparations by epidermal growth factor. *Journal of Biological Chemistry*. 254, 4884-4891.
[8] Tonks, N.K., Diltz, C.D. and Fischer, E.H. (1988). Characterization of the major protein-tyrosine-phosphatases of human placenta, *Journal of Biological Chemistry*. 263, 6731-6737.
[9] Resink, T.J., Hemmings, B.A., Tung, H.Y.L. and Cohen, P. (1983). Characterisation of a Reconstituted Mg-ATP-Dependent Protein Phosphatase. *European Journal of Biochemistry*. 133, 455-461.
[10] Barford, D., Flint, A.J. and Tonks, N.K. (1994). Crystal structure of human protein tyrosine phosphatase 1B. *Science*. 263, 1397.
[11] Picardi, P.K., Calegari, V.C., Prada, P.O., Moraes, J.C., Araújo, E., Marcondes, M.C.C.G., Ueno, M., Carvalheira, J.B.C., Velloso, L.A. and Saad, M.J.A. (2008). Reduction of hypothalamic protein tyrosine phosphatase improves insulin and leptin resistance in diet-induced obese rats. *Endocrinology*. 149, 3870-3880.
[12] Morrison, C.D., White, C.L., Wang, Z., Lee, S.-Y., Lawrence, D.S., Cefalu, W.T., Zhang, Z.-Y. and Gettys, T.W. (2007). Increased hypothalamic protein tyrosine phosphatase 1B contributes to leptin resistance with age. *Endocrinology*. 148, 433-440.
[13] Kaszubska, W., Falls, H.D., Schaefer, V.G., Haasch, D., Frost, L., Hessler, P., Kroeger, P.E., White, D.W., Jirousek, M.R. and Trevillyan, J.M. (2002). Protein tyrosine phosphatase 1B negatively regulates leptin signaling in a hypothalamic cell line, *Molecular and Cellular Endocrinology*. 195, 109-118.

[14] Zhang, S. and Zhang, Z.-Y. (2007). PTP1B as a drug target: recent developments in PTP1B inhibitor discovery. *Drug Discovery Today*. 12, 373-381.

[15] Cook, W.S. and Unger, R.H. (2002). Protein Tyrosine Phosphatase 1B: A Potential Leptin Resistance Factor of Obesity. *Developmental Cell*. 2, 385-387.

[16] Delibegovic, M., Zimmer, D., Kauffman, C., Rak, K., Hong, E.-G., Cho, Y.-R., Kim, J.K., Kahn, B.B., Neel, B.G. and Bence, K.K. (2009). Liver-specific deletion of protein-tyrosine phosphatase 1B (PTP1B) improves metabolic syndrome and attenuates diet-induced endoplasmic reticulum stress. *Diabetes*. 58, 590-599.

[17] Tonks NK, Diltz CD, Fischer EH. (1988) Purification of the major protein-tyrosine-phosphatases of human placenta. *J Biol Chem*.. 15; 263(14):6722-30.

[18] Ingebritsen, T.S. and Cohen, P. (1983). Protein phosphatases: properties and role in cellular regulation. *Science*. 221, 331.

[19] Charbonneau, H., Tonks, N.K., Kumar, S., Diltz, C.D., Harrylock, M., Cool, D.E., Krebs, E.G., Fischer, E.H. and Walsh, K.A. (1989). Human placenta protein-tyrosine-phosphatase: amino acid sequence and relationship to a family of receptor-like proteins. *Proceedings of the National Academy of Sciences of the United States of America*. 86, 5252-5256.

[20] Chan, R.J. and Feng, G.S. (2007). PTPN11 is the first identified proto-oncogene that encodes a tyrosine phosphatase. *Blood*. 109, 862-7.

[21] Ruddraraju, K.V. and Zhang, Z.-Y. (2017). Covalent inhibition of protein tyrosine phosphatases. *Molecular bioSystems*. 13, 1257-1279.

[22] Marsh-Armstrong, B., Fajnzylber, J.M., Korntner, S., Plaman, B.A. and Bishop, A.C. (2018). The Allosteric Site on SHP2's Protein Tyrosine Phosphatase Domain is Targetable with Druglike Small Molecules, *ACS Omega. 3*, 15763-15770.

[23] Singh, J., Petter, R.C. and Kluge, A.F. (2010). Targeted covalent drugs of the kinase family, *Current Opinion in Chemical Biology*. 14, 475-480.

[24] Wang, Q.P., Dechert, U., Jirik, F. and Withers, S.G. (1994). Suicide Inactivation of Human Prostatic Acid Phosphatase and a Phosphotyrosine Phosphatase. *Biochemical and Biophysical Research Communications*. 200, 577-583.

[25] Lo, L.-C., Wang, H.-Y. and Wang, Z.-J. (1999). Design and Synthesis of an Activity Probe for Protein Tyrosine Phosphatases. *Journal of the Chinese Chemical Society*. 46, 715-718.

[26] Jiang, J., Zeng, D. and Li, S. (2009). Photogenerated Quinone Methides as Protein Affinity Labeling Reagents. *ChemBioChem*. 10, 635-638.

[27] Scott W. Grimm, Heidi J. Einolf, Steven D. Hall, Kan He, et al. (2009). The Conduct of *in Vitro* Studies to Address Time-Dependent Inhibition of Drug-Metabolizing Enzymes: A Perspective of the Pharmaceutical Research and Manufacturers of America. *Drug Metabolism and Disposition.* 37 (7) 1355-70.

[28] Singh, J., Petter, R.C., Baillie, T.A. and Whitty, A. (2011). The resurgence of covalent drugs, *Nature Reviews Drug Discovery*. 10, 307.

[29] Grimm, S.W., Einolf, H.J., Hall, S.D., He, K., Lim, H.-K., Ling, K.-H.J., Lu, C., Nomeir, A.A., Seibert, E., Skordos, K.W., Tonn, G.R., Van Horn, R., Wang, R.W., Wong, Y.N., Yang, T.J. and Obach, R.S. (2009). The Conduct of *in Vitro* Studies to Address Time-Dependent Inhibition of Drug-Metabolizing Enzymes: A Perspective of the Pharmaceutical Research and Manufacturers of America. *Drug Metabolism and Disposition*. 37, 1355.

[30] den Hertog, J., Groen, A. and van der Wijk, T. (2005). Redox regulation of protein-tyrosine phosphatases. *Archives of Biochemistry and Biophysics*. 434, 11-15.

[31] Lee, S.-R., Yang, K.-S., Kwon, J., Lee, C., Jeong, W. and Rhee, S.G. (2002). Reversible Inactivation of the Tumor Suppressor PTEN by H2O2. *Journal of Biological Chemistry*. 277, 20336-20342.

[32] Barrett, W.C., DeGnore, J.P., König, S., Fales, H.M., Keng, Y.-F., Zhang, Z.-Y., Yim, M.B. and Chock, P.B., 1999. Regulation of PTP1B

via Glutathionylation of the Active Site Cysteine 215. *Biochemistry*. 38, 6699-6705.

[33] Bhattacharya, S., Labutti, J.N., Seiner, D.R. and Gates, K.S. (2008). Oxidative inactivation of protein tyrosine phosphatase 1B by organic hydroperoxides. *Bioorganic & medicinal chemistry letters*. 18, 5856-5859.

[34] Caselli, A., Camici, G., Manao, G., Moneti, G., Pazzagli, L., Cappugi, G. and Ramponi, G. (1994). Nitric oxide causes inactivation of the low molecular weight phosphotyrosine protein phosphatase. *Journal of Biological Chemistry*. 269, 24878-24882.

[35] Tanner, J.J., Parsons, Z.D., Cummings, A.H., Zhou, H. and Gates, K.S. (2010). Redox Regulation of Protein Tyrosine Phosphatases: Structural and Chemical Aspects. *Antioxidants & Redox Signaling*. 15, 77-97.

[36] Zhang, Y.-L., Yao, Z.-J., Sarmiento, M., Wu, L., Burke, T.R. and Zhang, Z.-Y. (2000). Thermodynamic Study of Ligand Binding to Protein-tyrosine Phosphatase 1B and Its Substrate-trapping Mutants. *Journal of Biological Chemistry*. 275, 34205-34212.

[37] Chen, L., Wu, L., Otaka, A., Smyth, M.S., Roller, P.P., Burke, T.R., Denhertog, J. and Zhang, Z.Y. (1995). Why Is Phosphonodifluoromethyl Phenylalanine a More Potent Inhibitory Moiety Than Phosphonomethyl Phenylalanine Toward Protein-Tyrosine Phosphatases. *Biochemical and Biophysical Research Communications*. 216, 976-984.

[38] Tulsi, N.S., Downey, A.M. and Cairo, C.W. (2010). A protected l-bromo-phosphonomethylphenylalanine amino acid derivative (BrPmp) for synthesis of irreversible protein tyrosine phosphatase inhibitors, *Bioorganic & Medicinal Chemistry*. 18, 8679-8686.

[39] Bradshaw, J.M., McFarland, J.M., Paavilainen, V.O., Bisconte, A., Tam, D., Phan, V.T., Romanov, S., Finkle, D., Shu, J., Patel, V., Ton, T., Li, X., Loughhead, D.G., Nunn, P.A., Karr, D.E., Gerritsen, M.E., Funk, J.O., Owens, T.D., Verner, E., Brameld, K.A., Hill, R.J., Goldstein, D.M. and Taunton, J. (2015). Prolonged and tunable

residence time using reversible covalent kinase inhibitors. *Nature chemical biology*. 11, 525-531.

[40] Liu, Q., Sabnis, Y., Zhao, Z., Zhang, T., Buhrlage, S.J., Jones, L.H. and Gray, N.S. (2013). Developing irreversible inhibitors of the protein kinase cysteinome. *Chemistry & biology*. 20, 146-159.

[41] Seiner, D.R., LaButti, J.N. and Gates, K.S. (2007). Kinetics and mechanism of protein tyrosine phosphatase 1B inactivation by acrolein. *Chemical research in toxicology*. 20, 1315-1320.

[42] Bradshaw, J.M., McFarland, J.M., Paavilainen, V.O., Bisconte, A., Tam, D., Phan, V.T., et al. (2015) Prolonged and tunable residence time using reversible covalent kinase inhibitors. *Nature chemical biology*.;11(7), 525-31.

[43] Shim, Y.S., Kim, K.C., Lee, K.A., Shrestha, S., Lee, K.-H., Kim, C.K. and Cho, H. (2005). Formylchromone derivatives as irreversible and selective inhibitors of human protein tyrosine phosphatase 1B. Kinetic and modeling studies. *Bioorganic & Medicinal Chemistry*. 13, 1325-1332.

[44] Arabaci, G., Yi, T., Fu, H., Porter, M.E., Beebe, K.D. and Pei, D. (2002). α-Bromoacetophenone derivatives as neutral protein tyrosine phosphatase inhibitors: structure–Activity relationship. *Bioorganic & Medicinal Chemistry Letters*. 12, 3047-3050.

[45] Lewis, S.M., Li, Y., Catalano, M.J., Laciak, A.R., Singh, H., Seiner, D.R., Reilly, T.J., Tanner, J.J. and Gates, K.S. (2015). Inactivation of protein tyrosine phosphatases by dietary isothiocyanates. *Bioorganic & Medicinal Chemistry Letters*. 25, 4549-4552.

[46] Zhang, Z.Y., Davis, J.P. and Van Etten, R.L. (1992). Covalent modification and active site-directed inactivation of a low molecular weight phosphotyrosyl protein phosphatase. *Biochemistry*. 31, 1701-1711.

BIOGRAPHICAL SKETCH

Marisa Cabeza

Present Position: Professor/Leader Researcher

Affiliation: Departamento de Sistemas Biológicos, Universidad Autónoma Metropolitana-Xochimilco

Marisa Cabeza has been researcher and professor at the Universidad Autónoma Metropolitana-Xochimilco since 1976. She was born in Mexico City, where she got her Ph.D degree in Basic Biomedical Research in 1995 at the Universidad Nacional Autónoma de México in Mexico City.

Since 1972 Professor Cabeza has been postgraduate professor in biochemistry, biomedical sciences, physiology and other disciplines related to medical sciences at the Universidad Nacional Autónoma de México and at the Universidad Autónoma Metropolitana. She began her research work as research assistant in the field of hormones in 1972 at the Instituto Nacional de Enfermedades de la Nutrición Salvador Zubirán, later to become head of research on steroidal hormones at the Universidad Autónoma Metropolitana. Her work has been recognized by Conacyt, which named her National Researcher level III in the National System of Researchers in Mexico.

She is the author of many publications in international journals, of five books, as well as of many book chapters related to hormones. She obtained a patent related to the production of anti-acne ointment that contains steroidal hormones. Dr. Cabeza has directed "licenciatura", master's and Ph.D theses at the Universidad Nacional Autónoma de México and at the Universidad Autónoma Metropolitana. In 1984 she was awarded the prize for the best contribution to Pharmacy by the Mexican Pharmaceutical Association. In 2003 she was awarded the prize for the best contribution to Urology and Nephrology by the National Hospital of México. In 2011 she was prized by the best contribution for research on steroids hormones" was granted to her in the Congress of Research of Steroids with the work "New ester derivatives of dehydroepiandrosterone as inhibitors of 5α-reductase". Awarded by Elsevier. United States. In 2018 she was awarded by the best professor of the year by Universidad Autónoma Metropolitana-Xochimilco.

She belongs to the Western Pharmacology Society and to the Asociación Mexicana de Bioquímica Clínica, Endocrine Society. Now she is a member of the Biological Systems Department at the Universidad Autónoma Metropolitana-Xochmilco as full professor and researcher, and has focused her research in the field of the biological effects of new steroidal compounds which have been synthesized by the group which Dr. Cabeza co-directs. This research has the aim to find molecules with potential therapeutical use as antiandrogens, to improve diseases such as benign prostatic hypertrophy and prostate cancer.

Education:

- BS degree in Biology 1973. National University of Mexico City. Thesis: *In vitro* metabolism of [3H] Testosterone in the hypothalamus, hypocampus and pituitary gland of castrated rats.
- MS degree in reproductive biology 1991. National University of Mexico City. Thesis: Effect of testosterone and levonorgestrel on the lipid synthesis of [$U^{14}C$] glucose in female hamster flank organs.
- PhD degree in Biomedical Research 1995. National University of Mexico City. Thesis: Molecular interaction of levonorgestrel and its 5-alfa-metabolite with androgen receptors in hamster flank organs.

Research and Professional Experience:

1984-present. Full professor of Biomedical Research, Cellular Processes, Biochemistry, Pharmacology, Medicinal Chemistry in the Metropolitan University of Mexico City. Investigation project at the University: Effect of hormones on different tissues. I am teaching the following post-graduate courses:

1. Health and population: Biomedical Bases of Family Planning.
2. Drugs Organism Interaction
3. Biological Science

Publications from the Last 3 Years:

1. López-Lezama J, Soriano-Garcia M, Valencia-Islas NA, Cabeza M. Crystal Structure and Synthesis of N-cyclohexyl-3β-hydroxyandrost-5,16-diene-17-carboxamide dihydrate. *X-ray Structure Analysis Online.* 2014, 30(12), 59. ONLINE ISSN 1883-3578
2. Cabeza M, Heuze Y, Sánchez A, Garrido M, Bratoeff E. Recent advances in structure of progestins and their binding to progesterone receptors. *Journal of Enzyme Inhibition and Medicinal Chemistry* 2015, 30(1): 152-159, DOI: 10.3109/14756366.2014.895719. ISNN: 1475-6366 (print) 1475-6374 (online)
3. Garrido M, González-Areanas A, Camacho-Arroyo I, Cabeza M, Alcaráz B, Bratoeff E. Effect of new hybrids base don 5,16-pregnadiene scaffold linked to an anti-inflamatory drug on the growth of human astrocytoma cell line (U373). *European Journal of Medicinal Chemistry* 2015, 93: 135-141.ISNN: 0.023-5234 DOI: 10.1016/j.ejmech.2015.01.048
4. Ortiz AVS., Cabeza M, Bratoeff E, and Soriano-García M.Crystal Structure and Synthesis of Two Steroidal Derivatives: 3β-Propanoyloxyandrost-5-en-17-one and 3β-Pentanoyloxyandrost-5-en-17-one. *X-ray Structure Analysis Online*, 2015, 31(2), 9.
5. Chávez-Riveros A, Bratoeff E, Heuze Y, Soriano J, Moreno I, Sánchez-Márquez A, and Cabeza M. Synthesis and Identification of Pregnenolone Derivatives as Inhibitors of Isozymes of 5α-Reductase. *Arch. Pharm. Chem. Life Sci.* 2015, 348 (11), 808-816. ISNN: 1521-4184 DOI: 10.1002/ardp.201500220
6. Cabeza M, Medina Y, Álvarez B, Moreno I, Bratoeff E. Biological activity of novel 17β-phenylcarbamoyl-androst-4-en-3-one as inhibitors of type 2 5α-reductase enzyme. *Memorias de Congreso*

publicadas por: Athens Institute for Education Research, Paper Series N°: PHA2015-1446. [*Memoirs of Congress* published by: Athens Institute for Education Research]

7. Silva-Ortiz A V, Bratoeff E †, Ramírez-Apan T, Heuze Y, Sánchez A, Soriano J, Cabeza M,.Synthesis and activity of novel 16-dehydropregnenolone acetate derivatives as inhibitors of type 1 5α-reductase and on cancer cell line SK-LU-1 in *Bioorganic & Medicinal Chemistry* 2015,23, 7535-7543.DOI: 10.1016/j.bmc. 2015.10.047
8. .Cabeza M, Posada A, Sánchez-Márquez A, Heuze Y, Moreno I, Soriano J, Garrido M, Cortés F, and Bratoeff E Biological activity of pyrazole and imidazole-dehydroepiandrosterone derivatives on the activity of 17β-hydroxysteroid dehydrogenase. *J. Enzyme Inhibition and Medicinal Chemistry* 2016, 31(1):53-62. ISNN: 1474-6366 (PRINT), ELECTRONIC 1475-6374. ID: 1003926 DOI:10.3109/14756366.2014.1003926
9. Arellano Y, Bratoeff E, Segura T, Mendoza ME, Sánchez-Márquez, Medina Y, Heuze Y, Soriano J, and Cabeza C. Novel dehydroepiandrosterone benzimidazolyl derivatives as 5α-reductase isozymes inhibitors. *J. Enzyme Inhibition and Medicinal Chemistry*, 2016, 31 (6)908-914. DOI: 10.3109/14756366.2015.1070843
10. Sánchez-Márquez A, Silva-Ortíz A, Bratoeff E, Heuze Y, Soriano J, Medina Y and Cabeza M. New Dehydroepiandrosterone-triazole Derivatives Identified as Inhibitors of 17β-Hydroxysteroid Dehydrogenase Enzyme in the Prostate. *Current Enzyme Inhibition*, 2016, 12(2) 145-154. ISNN (Print) 1573-40-80. (Online 1875-6662)
11. Cabeza M, Sánchez-Márquez A, Garrido M, Silva A, Bratoeff E. Recent Advances in Drug Design and Drug Discovery for Androgen-Dependent Diseases. *Current Medicinal Chemistry* 2016, 23(8) 792-815. ISSN (Print): 0929-8673 ISSN (Online): 1875-533X - BRRk9qtZ.dpuf DOI: 10.2174/0929867323666160210125642
12. .Sánchez-Márquez A, Arellano Y, Bratoeff E, Heuze Y, Córdova K, Nieves G, Soriano J & Cabeza M. Synthesis and biological evaluation of esters of 16-formyl-17-methoxy-dehydro-

epiandrosterone derivatives as inhibitors of 5α-reductase type 2. *J. Enzyme Inhibition and Medicinal Chemistry*, 2016 31(6) 1170-1176. DOI:10.3109/14756366.2015.1103235. ISNN: 1475-6366

13. Cortés-Benítes F, Cabeza M, Ramírez-Apan MT, Álvarez-Manrique B, Bratoeff E. Synthesis of 17β-N-arylcarbamoylandrost-4-en-3-one derivatives, and their anti-proliferative effect on human androgen-sensitive LNCaP cell line. *European Journal of Medicinal Chemistry* 2016, 121:737-746. http://dx.doi.org/10.1016/j.ejmech.2016.05.059
14. Silva-Ortiz AV, Bratoeff E, Ramírez-Apan T, Heuze Y, Soriano J, Moreno I, Bravo M, Bautista L, Cabeza M. Synthesis of new derivatives of 21-imidazolyl-16-dehydropregnenoloneas inhibitors of 5α-reductase 2 and with cytotoxic activity in cancer cells. *Bioorganic and Medicinal Chemistry* 2017, 25: 1600-1607. ISNN (Print) 0968. (Online 0896) http://dx.doi.org/10.1016/j.bmc.2017.01.018
15. Bratoeff E, Moreno I, Cortés-Benítez F, Heuze Y, Bravo M, and Cabeza M. 17β-N-arylcarbamoylandrost-4-en-3-one Derivatives as Inhibitors of the Enzymes 3α-Hydroxysteroid Dehydrogenase and 5α-Reductase. *Current Enzyme Inhibition* 2018, 14:36-50. 15:1-15.ISNN (Print) 1573-4080. (Online 1875-6662)
16. Arellano Y, Bratoeff E, Heuze Y, Bravo M, Soriano J. and Cabeza M. Activity of Steroid 4 and derivatives 4a-4f as inhibitors of the Enzyme 5α-reductase 1. *Bioorganic and Medicinal Chemistry* 2018, 26:4058-4084. ISNN (Print) 0968. (Online 0896) https://doi.org/10.1016/j.bmc.2018.06.030
17. Cabeza M, Bautista L, Bravo MG; Heuze Y. Molecular Interactions of Different Steroids Contributing to Sebum Production. Review. *Current Drug Targets*. 2018, 19 (15) 1855-1865. ISNN (Print) 1389-4501. https://doi.org/10.2174/1389450119666180808113951

INDEX

B

C

D

E

F

G

H

I

K

L

M

N

O

P

R

S

T

U

V

W

Y

Z

α

Related Nova Publications

Calmodulin: Structure, Mechanisms and Functions

Editor: Vahid Ohme

Series: Cell Biology Research Progress

Book Description: In *Calmodulin: Structure, Mechanisms and Functions,* the authors consider small and poorly-studied groups of plant calcium-dependent protein kinases that directly interact with calmodulin molecules.

Softcover ISBN: 978-1-53614-948-7
Retail Price: $82

Flagella and Cilia: Types, Structure and Functions

Editor: Rustem E. Uzbekov

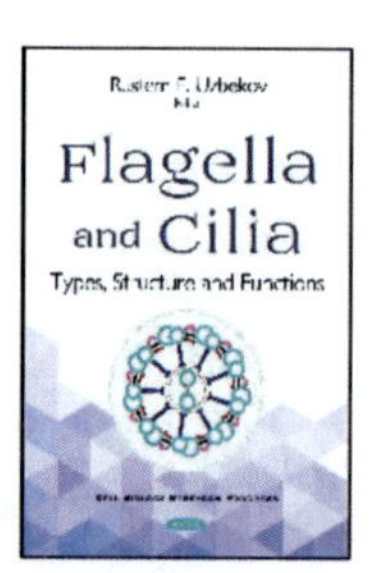

Series: Cell Biology Research Progress

Book Description: Motility is an inherent property of living organisms, both unicellular and multicellular. One of the principal mechanisms of cell motility is the use of peculiar biological engines – flagella and cilia. These types of movers already appear in prokaryotic cells. However, despite the similar function, bacteria flagellum and eukaryote flagella have fundamentally different structures.

Softcover ISBN: 978-1-53614-333-1
Retail Price: $95

To see a complete list of Nova publications, please visit our website at www.novapublishers.com